Gender and Religion in the City

This book provides a conceptual, historical and contemporary context to the relationships between gender, religion and cities.

It draws together these three components to provide an innovative view of how religion and gender interact and affect urban form and city planning. While there have been many books that deal with religion and cities; gender and cities; and gender and religion, this book is unique in bringing these three subjects together. This trio of inter-relationships is first explored within Western Christianity: in Roman Catholicism, Protestantism, Eastern Orthodoxy and in the Pentecostal and Charismatic movements. A wider perspective is then provided in chapters on the ways in which Islam shapes urban development and influences the position of Muslim women in urban space. While official religions have declined in the West there is still a desire for new forms of spirituality, and this is discussed in chapters on municipal spirituality and on the rise of paganism and the links to both environmentalism and feminism. Finally, ways of taking into account both gender and religion within the statutory urban planning system are presented.

This book will be of great interest to those researching environment and gender, urban planning and sustainability, human geography and religion.

Clara Greed is Emerita Professor of Inclusive Urban Planning at the University of the West of England, Bristol, UK. She is the author/editor of over 12 books, including, *Women and Planning: Creating Gendered Realities* (1994), *Planning in the UK: An Introduction* (2014), *Inclusive Urban Design: Public Toilets* (Routledge, 2003).

Routledge Studies in Gender and Environments

With the European Union, United Nations, UN Framework Convention on Climate Change, and national governments and businesses at least ostensibly paying more attention to gender, including as it relates to environments, there is more need than ever for existing and future scholars, policy makers, and environmental professionals to understand and be able to apply these concepts to work towards greater gender equality in and for a sustainable world.

Comprising edited collections, monographs and textbooks, this new *Routledge Studies in Gender and Environments* series will incorporate sophisticated critiques and theorisations, including engaging with the full range of masculinities and femininities, intersectionality, and LBGTIQ perspectives. The concept of 'environment' will also be drawn broadly to recognise how built, social and natural environments intersect with and influence each other. Contributions will also be sought from global regions and contexts which are not yet well represented in gender and environments literature, in particular Russia, the Middle East, and China, as well as other East Asian countries such as Japan and Korea.

Martin Hultman is a Senior Lecturer at Linköping University, Sweden.

Virginie Le Masson is a Research Fellow at the Overseas Development Institute, London, UK.

Sherilyn MacGregor is a Reader in Environmental Politics at the University of Manchester, UK.

Tanja Mölders is an environmental scientist. Since 2013 she is University Professor and holds the chair 'Space and Gender' at Leibniz University Hannover, Germany.

Karen Morrow is Professor of Environmental Law at Swansea University, UK.

Marion Roberts is Professor of Urban Design at Westminster University, UK.

Titles in this series include:

Ecological Masculinities
Theoretical Foundations and Practical Guidance
Martin Hultman and Paul M. Pulé

Gendered Approaches to Spatial Development in Europe
Perspectives, Similarities, Differences
Edited by Barbara Zibell, Doris Damyanovic and Ulrike Sturm

Gender and Religion in the City
Women, Urban Planning and Spirituality
Edited by Clara Greed

Gender and Religion in the City

Women, Urban Planning and Spirituality

Edited by Clara Greed

Routledge
Taylor & Francis Group
LONDON AND NEW YORK

earthscan
from Routledge

First published 2020

by Routledge
2 Park Square, Milton Park, Abingdon, Oxon OX14 4RN
and by Routledge

52 Vanderbilt Avenue, New York, NY 10017

Routledge is an imprint of the Taylor & Francis Group, an informa business

First issued in paperback 2021

British Library Cataloguing-in-Publication Data
A catalogue record for this book is available from the British Library

Library of Congress Cataloging-in-Publication Data
A catalog record has been requested for this book

ISBN: 978-1-138-38510-8 (hbk)
ISBN: 978-1-03-208534-0 (pbk)
ISBN: 978-0-429-42733-6 (ebk)

Typeset in Goudy
by Wearset Ltd, Boldon, Tyne and Wear

Contents

Figures

Tables

Contributors

Mokarram Abbas is an architect from Nablus, Palestine. She undertook her PhD in Belgium at the Vrije Universiteit (Free University), Brussels, supervised by Dr Bas van Heur. She has worked in the Department of Architecture and Urban Design at the Catholic University in Leuven, Belgium and undertook urban design research at the University of Westminster. Her research has focused upon the spatial position of women in Muslim cities, and the question of whether the Western 'women and planning' script is applicable in Muslim cities. On returning to Palestine, Mokarram became Dean of the Faculty of Engineering at Palestine Technical University, was elected city councillor in Tulkarm in the West Bank, and also started a family. So, she has little time to publish but her work was featured in *Gender Place and Culture*.

Chris Baker is Professor of Religion and Public Life, at Goldsmiths, University of London and was previously in the theology department at Chester University. At Goldsmiths he is attached to the Faiths and Civil Society Unit and the Department of Social, Therapeutic and Community Studies. Chris is also Director of Research for the William Temple Foundation. His research and publications explore the impact of embodied religion and belief on public life in the modern world. His research covers such areas as community development, local policy, professional public practice, urban development and planning policy, as well as business and work-based leadership and management, and new forms of political and civic activism. He is a key figure in the development of postsecular urban thought and has published widely on postsecular cities.

Juliet Davis is Reader in Architecture and Urbanism at the Welsh School of Architecture at Cardiff University. She originally studied Architecture at Cambridge University and practised in London for Eric Parry Architects for six years before entering academia. Projects included new studios for the Wimbledon School of Art and a social care facility for St Martin in the Fields. She completed her PhD on Planning for London's Olympic urban regeneration legacy at the London School of Economics' Cities Programme in 2011. Recent publications include articles in *Planning Perspectives* and the

Journal of Urbanism on urban change in East London neighbourhoods. She is currently working on a book that draws on the ethics of care literature to discuss issues of inclusive urbanism, which highlights women as the main carers in society).

Carey Doyle is a chartered town planner and member of the Institute for Environmental Management and Assessment. She has worked as a practising planner in the UK for over 14 years, and holds a Master's degree from the University of Chicago. She completed her doctoral research in 2018 at Queen's University Belfast on planning and migration. This research identified the ways in which current planning practice does not address diversity, or the implications of migration for planning's redistributive functions. In 2017 she led a major study on planning for growth in religious communities for the London Borough of Barking and Dagenham, for CAG, the planning consultancy. This identified existing and future needs for religious meeting places, advised the local council on consultation with faith groups and addressed the requirements for religious facilities in new housing developments.

Clara Greed, the editor, is Professor Emerita of Inclusive Urban Planning at the University of the West of England, Bristol. Originally a town planner, her main research has been on gender, equality and accessibility in the city of man. She has increasingly been concerned not only with the nature of urban spatial planning, but quite 'who' is making the policy decisions, as shaped by the planning profession's subculture, which is generally male-orientated and also secular in perspective. Clara is motivated in all her work by 'planning for everyday life', which is why she has concentrated on such diverse issues as public toilet provision, and increasingly the role of religion and faith in cities: both factors below the radar of the planners but of great concern to many ordinary people.

Carmen Hass-Klau is an internationally recognised transport planner, known for her work on traffic calming and planning for the pedestrian, as set out in *The Pedestrian and the City* (2015). Carmen qualified as an engineer at the Technical University of Berlin. She took her PhD at Reading University in transport planning under the supervision of Professor Peter Hall. Until 2012 she was Professor of Civil Engineering: Transport and Public Transport Systems in Europe at the University of Wuppertal, Germany. She has undertaken consultancy work on transport planning for over 30 years and continues to do so. She is currently investigating how urban form is shaped by religion and culture, identifying the differences between historically Christian and Muslim cities and exploring the question of which of these gives greater priority to the pedestrian.

Petra Kuppinger is Professor of Anthropology in the Department of Sociology/Anthropology at Monmouth College, Monmouth, IL, USA. Her main interests are in cities and spaces; space and power; Middle Eastern cities, space culture and Islam in Germany; globalisation and globalised consumer cultures. She has undertaken extensive research in Germany on the relationship between North European Protestants, especially Lutheran churches, and Islamic German faith groups as evidenced in her chapter. She is editor of the journal *City and Society* and was President of the Society of Urban National and Transnational Anthropology (2014–2016). Her recent publications include 'Flexible Topographies: Muslim Spaces in a German Cityscape', *Social and Cultural Geography* (2014). She edited with George Gmelch the 2017 edition of *Urban Life: Reading in the Anthropology of the City*.

Katie McClymont is a graduate of the PPE (Philosophy, Politics and Economics) degree at Cambridge University. She undertook her PhD at Sheffield University Town Planning Department. She is a member of the Planning team at the University of the West of England, Bristol. Katie has been researching the attitudes of local planning authorities towards religious buildings and spaces and the churches' role in the community. Such 'municipal spirituality' is 'beneath the radar' of many planners, maybe because women are the volunteers. She is also researching the relationship between planning and death. She undertook an AHRC project entitled *Deathscapes and Diversity in multicultural England and Wales* and is working on a British Academy Rising Star Engagement Award entitled *Spaces for Secular Faith: Planning Practice, Shared Assets and Intangible Values in Diverse and Changing Communities*.

Andrea Mulkeen is a planning professional who works for the Church of England, with the Church Commissioners, managing their properties and estates and dealing with the planners. She has produced reports on ecclesiastical buildings in association with the Church Buildings Review Group on behalf of Historic England. She has also undertaken research for the Church Buildings Review Consultation, the Institute of Historic Building Conservation, and the Historic Religious Buildings Alliance (HRBA), which is a group within the Heritage Alliance which brings together those working for a secure future for historic religious buildings. She produced a report for the University of Westminster in 2014 entitled *The Planning System and Places of Worship: The Extent to Which the Planning System since the Second World War Has Facilitated the Needs of Faith Communities for Places of Worship*.

Anna Papadopoulou gained her doctorate at Cardiff University in 2018 under the supervision of her co-author of Chapter 5, Juliet Davis. Her research focused upon investigating the position of women in Limassol in Cyprus, with reference to both the historical influence of the Greek Orthodox Church, and the role of modern town planning policy, and in shaping city form and women's lives therein. Anna brings an international perspective to her investigations as she also holds a Master's degree in Landscape Architecture from the Rhode Island School of Design, and a Post-Baccalaureate diploma in Urban Studies and a Bachelor of Science in Geological Sciences and Classics from Tufts University. Anna is an adjunct in the Faculty in Architecture at the University of Nicosia and University of Cyprus, where she teaches architectural design.

Gill Reeve is an ordained pioneer minister in the Church of England, working in the progressive and inclusive parish of St Luke's in the City, Liverpool. She is also studying for a PhD at Goldsmiths, under Chris Baker's supervision, on the place of religious initiatives in urban governance networks. Gill's background is as a speech and language therapist, working in the health service for 15 years, which was foundational to her commitment to community engagement initiatives, collaborative partnership working and social inclusion. In 2010, as Director of a small Christian charity she pioneered 'Night Church', an inclusive Christian sacred space in the night time economy of a northern UK city. She led the project until 2016 when she left to begin ordination training, which continues under the leadership of a member of Gill's original team.

Inés Sánchez de Madariaga is Chair on Gender and Professor of Urban Planning at Madrid School of Architecture. She has been a visiting professor at the London School of Economics, at the Bauhaus School of Architecture and at Columbia University, New York, where she obtained her MSc as a Fulbright scholar. She is Director of the UNESCO Chair on Gender, and an advisor to many international bodies. She has been chair of the EU-funded COST Gender STE programme and Co-Director of the EU-US Gendered Innovations Project. In Spain, she has been Director of the Women and Science Unit at the Cabinet of the Minister of Science and Innovation, and Deputy Director for Architecture, and Member of the Cabinet of the Minister of Housing. She has published widely on gender and cities, including editing, with Marion Roberts, the collection, *Fair Shared Cities* (2013).

Roja Tafaroji is from an Iranian Muslim background, now living in Auckland, New Zealand. Roja's doctoral research, initially supervised by Dory Reeves, focuses upon Afghani Muslim women immigrants who have settled in Auckland. She is investigating how they relate to the spatial and social setting of the city, with reference to the neighbourhoods in which they live, the journeys they make, and how their experiences differ from male Afghani immigrants. Roja obtained her BA in Architecture from Guilan University in Iran and her Master's in Urban Design from Iran University of Science and Technology (IUST). Her Master's thesis explored the main factors which enable women to enter public spaces with reference to a section of Vali-e-Asr Street in Tehran. She has presented her research at STE and AESOP conferences and continues with her research.

Preface

While there has been a great deal of publication on the duos of 'gender and religion', 'religion and cities' and 'gender and cities', there has been far less research on how the trio of religion, gender and cities interrelate. This book comprises a collection of 15 chapters from different perspectives, and seeks to investigate the relationships and interactions between religion (faith, spirituality), gender (and women) and cities (and thus urban planning). Religion is a topic that is often missing from the urban planner's view of the city, and therefore is generally absent from spatial policy making. Yet, historically, religion has been a major factor in shaping city form, urban design and architecture. Likewise, gender, and especially the consideration of women's needs within the city of man, has, until relatively recently, been absent from the planning agenda. Yet religion and women's issues exist in an uneasy relationship. Feminism, and gender studies, are often characterised as secular and humanistic in perspective, with religion being seen as highly patriarchal and reactionary. Nevertheless, there is arguably a renewed interest in religion, as Western society moves from a secular to a postsecular phase, and women's issues often figure strongly within this new agenda. There has not only been an advance of Islam in the West, but also growth and renewal within certain sectors of Christianity, and a general societal curiosity in generic spirituality, paganism and non-traditional religions.

Following a contextualising introduction to the concept of the postsecular city, and its implications for modern society, four chapters discuss the relationship between gender, religion and urban form, with reference to the main forms of Christianity, namely Roman Catholicism, Eastern Orthodoxy, Protestantism, but also the Pentecostal and Charismatic movements. The spatial dimensions will be illustrated especially in relation to urban form and local design. It is clear from these chapters that there is no one simple 'image of the city' as often assumed by male urban design pundits, but that women's experience of the city is often very different from men's, with a completely different set of key nodal points, access route ways and urban spatial demarcations and delineations.

With increasing globalisation, immigration and inter-faith dialogue it is impossible to discuss any aspect of Christianity in a watertight compartment.

Also Christianity is a 'very broad church' and it will be seen that the authors hold some quite different viewpoints, especially on the role of women in the Church. A key question – and a rather non-politically correct one at that – which arises from these chapters is to what extent the different roles that churches often assign to men as against women, actually restrict women or enable them to achieve a significant level of autonomy and participation in society and urban space.

The next section looks at Islam. An introductory chapter sets out the differences between historical Christian and Muslim urban form. Next the modern-day relationship between women and urban form is discussed with reference to a predominantly Muslim city, Nablus in Palestine, where gendered spatial divisions appear impregnable but are actually complex, multilayered and evolving. Then the situation in the multi-national global 'Western' city of Auckland in New Zealand is illustrated from the perspective of Muslim women immigrants, where they have increased freedom but traditions from the home country remain strong and inform their choice of location and how they relate to the local neighbourhood. Discussing the position of women within the Muslim city raises questions as to whether 'separation' of men and women in urban space is always 'bad for women' and whether 'mixing' of the sexes brings with it its own challenges for Muslim women but also for women in Western cities where gender neutrality is increasingly being promoted and traditional women-only spaces are being challenged as out of date.

Following this, two chapters discuss how women's lives and urban needs are affected by new forms of religion (or none), and the concept of municipal spirituality. Formal historical religions are in decline but there is still a desire for specific urban spaces and buildings for reflection, memorial, gatherings, community and prayer. The relationship between spirituality, paganism, environmentalism and feminism is also investigated. The final section addresses the relationship between urban planning and religion. It recounts how religion was gradually 'streamlined' out of the UK statutory planning system in the second part of the twentieth century. Then the role of community-related planning consultancy and advocacy in promoting the importance of both religion and gender is illustrated. Drawing on the case study of the planning problems KICC Church (a Black, Pentecostal, London-based mega church) experienced are presented, along with an explanation of the legal and procedural problems created by the different levels and policy specialisms of the planning system. Finally, a means of ameliorating the situation for the future is presented, namely a Religion Mainstreaming Toolkit that can be used both in relation to urban policy making and in determining planning applications, The aim is to enable 'religion' to be included in planning practice, just like any other land use or policy priority along with all the other 'normal' economic, social, environmental and spatial factors involved. It is hoped that this book will raise awareness, amongst planners, of the role of religion in generating urban form, while still taking on board the importance of gender. But all along we acknowledge that the situation is complex, and we do not minimise all the contradictions, differences of view and unresolved issues inherent in investigation of the relationship between religion, gender, cities and urban planning.

Clara Greed
Bristol

Part I

Background

1 Introduction

Interactions between gender, urban space and religion

Clara Greed

How can the humanities ignore the human soul?

(Arthur, 1991)

Introduction

This book investigates the relationship between religion (including faith, belief, spirituality, worship), cities (and urban planning) and gender (with particular reference to women). In our predominantly secular, material and scientific culture, 'religion' is often seen as an outdated concept, a superstition with little relevance to the modern age. But for millennia religious belief has had a major influence in shaping urban areas, both in the UK and globally. Religion has also shaped women's position within society and the city, in both negative and positive ways. There has always been an uneasy relationship between religion and feminism because of the perceived patriarchal and misogynist nature of the Church and its apparent endorsement of women's oppression within society (Daly, 1973). Likewise the built-environment professions have been male-dominated and have historically provided little space for the consideration of women's needs within the city of man (Greed, 1994).

There is a great deal of research and publication on the duos of 'gender and religion', 'gender and cities' and 'religion and cities' (for example Eade and Mele, 2002), but this book seeks innovatively to achieve the hat-trick (trio), to combine all three factors together (as pre-figured in Greed, 2011). But we are hindered by the 'double blindness' of both religion and gender seldom figuring in either planning policy or urban sociology (Jaschok, 2015; Greed, 2017) or in the humanities as a whole (as commented upon in the statement at the top of this chapter) (Arthur, 1991). Therefore, throughout, we will inform our discussions of religion and cities with a gendered lens on the issues.

The situation is not static. It has been argued that society is now, in part at least, entering a postsecular phase (Beaumont and Baker, 2011), characterised by a re-emergence of faith in the public realm of civil society and the city (Cox, 1995; Beaumont and Baker, 2011), compared with twentieth-century attitudes (Cox, 1965). But many traditional Christians are not convinced and believe we are still in the secular phase (Cornwell, 2019). As to definitions of 'postsecularism',

Baines argues that, according to Habermas, postsecularism is marked by a contest between religious lobbyists (which have grown in strength with the resurgence of fundamentalist religions) and secular pressure groups who promote diversity and equality (Baines, 2009). A new relativism, and an uneasiness with scientific positivism, are also features of the emerging postsecular era, allowing for the co-existence of a range of truths, with no one truth predominating which is at odds with both secular scientific positivism and the absolutes of sacred dogma (Habermas and Ratzinger, 2007). Within this evolving context, women and other marginalised groups have been able to challenge the principles of religious orthodoxy from a feminist perspective. This has enabled women both to have a greater role within existing denominations (for example as vicars in the Anglican Church) and to create alternative ways of 'doing religion' often linked to creating new forms of utopian community.

Although this book is set within the context of postsecularism, it should not be assumed that people are returning to traditional Christianity. Many people, who would not consider that they have any particular religious affiliation, especially not to the Church, nevertheless have a longing for spirituality, reflection and contemplation (Davie, 1994; 2015). In the UK there has been a growth in the importance of Islam (Peach and Gale, 2003) along with the manifestation of a variety of new non-traditional Christian movements. These include Pentecostalism and the Charismatic movement (characterised by the emphasis upon the Holy Spirit (Burgess and van der Maas, 2002)). There is also an interest in paganism, and new age spirituality, some of whose adherents have sought to create alternative ways of living, often strongly espousing green, sustainable, environmental principles too, but who are also concerned about gender and sexualities (Kraemer, 2012).

Historically, religion has been a major factor in shaping civilisation, music, art, architecture and town planning, and the wider countryside and environment (Gorringe, 2002; Sheldrake, 2014; Garbin and Strhan, 2017). In the past urban planners were fully cognisant of its importance and designed entire cities around religious principles, for example during the Renaissance in Europe or in the Moghul and Hindu cities of India (Narayanan, 2016). In the UK, this religious heritage, the nation's 'inherited religious capital' of the Christian past, arguably still shapes modern secular society. Societal concerns with ethics, justice, human rights, politics, social engagement and equality issues of all sorts (and of course feminism) are all influenced, or at least framed, to some degree at least, by previous religious attitudes. But these roots have been obscured by secular and scientific ways of thinking. In contrast much of the rest of the world never underwent a secular phase, and religion still plays an important part in people's daily lives. When immigrants settle in the UK they bring with them their religious beliefs and related spatial needs to the bemusement of secularised urban planning authorities (Gale, 2008; Mee, 2016).

The spatial dimensions of religion

Several of the contributors to the book are urban planners (or at least from the realms of 'gender and geography' or linked to the 'women and planning' movement), and we are keen to give a strongly spatial perspective on the issues and to consider how the planning system treats 'religion'. But many mainstream planners seem to give very little 'value' to religious buildings, land uses and spaces, and there is no mention at all of religion in a typical planning department's development plans (Greed, 2016a; 2016b; Greed and Johnson, 2014). As will be explained, churches and other religious uses were once required to be included as part of the land-use planning system created after the Second World War. But by the 1970s this aspect was gradually reduced and following a 'streamlining' of planning law a separate religious category was dropped altogether (Mulkeen, 2014). At best nowadays religious buildings may be included under the category of diversity, culture or community, but at worst they may be left out altogether. Planners seem keen to treat religious buildings as disposable items or as 'heritage architecture', especially 'redundant' church buildings. Then they can be used for (what were seen as) 'more useful purposes' such as wine bars, apartments, museums, tourist attractions and arts centres – even when there is a great demand from other religious groups for redundant church buildings. Religious land use, development and building stock requirements are seldom recorded or surveyed by the planners, so they are rendered invisible. Therefore an acknowledgement of the 'spatial' dimensions of the religious and spiritual dimensions of people's lives and longings will suffuse much of the book.

But identifying the spatial dimensions of religion, and the need to integrate this into the planning system, does not necessarily mean that change will take place. We are living within a built environment that has taken centuries to develop, and that is not quickly altered. However, contributors can at least highlight 'the problems' and some will make recommendations as to how both urban policy and planning law might be reconfigured to improve the situation. Another key theme that is explored in many chapters is whether those who want to change the city and urban society (including feminists, religious groups and reformers) can achieve this through the planning system, from 'within' the planning system, or whether they chose to try to 'start again' by giving up on existing cities and planning systems and building utopian alternative communities instead. Both these approaches (trying to change existing cities or creating new alternatives) are based upon the theological assumption that Christian 'salvation' is not just something that is experienced when we die and go to Heaven (to the New Jerusalem, the ultimate heavenly Utopia). Rather it is the believer's duty to help fulfil the words of the Lord's Prayer, 'Thy Kingdom come on earth, as it is in Heaven' (Matthew Ch. 6, Ver. 10): that is faith and works.

The gendered aspects of religion

There has always been an uneasy relationship between religion and feminism because of the perceived patriarchal and misogynist nature of the Church and its apparent endorsement of women's oppression within society. Many religious women have problems with theologians and church leaders, particularly those on the right. In spite of assuring us of their best intentions, and that 'the Gospel is for everyone', some male leaders still manifest very traditional and often quite misogynistic attitudes as to quite who 'everyone' is. While it is to be welcomed that churches are getting more involved in 'the public square' and politics and involving 'more women' in the process (Boot, 2014; Laurence, 2014), it can be alarming that some more right-wing fundamentalist groups bring with them quite traditional attitudes towards women. There is always a tension between religion and gender, and, for example, individual women often feel they should remain faithful to their religious fellowship in order to do all their good works and worship their God. But, they may feel they are undervalued and undermined by the male leadership. Instead, they might find their vocation is better fulfilled by working in secular and governmental agencies.

Debates about secular and postsecular society leave many women unimpressed, because gender issues have not necessarily been acknowledged within the discourse (Aune, 2015). Likewise, we are not going to pretend that the inclusion of religion in planning policy will necessarily make cities better for everyone, especially women. Furthermore, many women are wary of patriarchal religion. The built-environment professions remain male-dominated and any discussion or inclusion of religion in the planning agenda is likely to be prone to be framed within the terms of male interests and patriarchal images of religion, unless mitigated by a gender perspective. One also needs to investigate aspects of feminist thought, for much feminism is still quite secular and humanistic in perspective. Some proponents of the latest fourth wave feminism seem quite hostile to religion and prone to accept negative stereotypes of religion (Greed, 2017). So faith still sits uneasily with feminism. Much of feminist thought is not particularly 'spatial' either. Certain branches of academic feminism tend to be rather 'snooty' about anything to do with buildings, construction, urban planning and the practicalities of life (Greed, 1991), preferring to devote their time to philosophy, sociology and cultural studies. This is because they may not appreciate the importance of the built environment, or simply take it for granted, for as David Harvey commented many years ago, many sociologists [and feminists] are 'living in a space-less vacuum' (Harvey, 1975).

However, the links between religion, cities and women's rights were more strongly manifest in first wave feminism but were gradually forgotten (Greed, 1994). Although women had been influential in the nineteenth-century urban reform movements and in the development of model utopian industrial communities, they were increasingly excluded from the new planning profession. Significant numbers of educated women actively campaigned to get women's needs and 'different' perspective incorporated into housing design, model town development, urban planning, architecture and in welfare reform. Many first wave feminists, including Charlotte Perkins Gilman, Olive Schreiner and Alice Constance Austin, were involved in town planning, housing and architecture (Greed, 1994: Chapter 6). In spite of the gendered restrictions present in Victorian Church-dominated society, many early feminists were also religiously inclined. Many had a strong moral agenda and were supporters of the temperance movement and Sunday observance. For example, a million women signed a petition in 1887 to make pubs close on Sundays (so the men did not drink their wages). The full rallying cry of the suffragette movement was 'votes for women and chastity for men' (which tied in with their three main principles of purity, dignity and hope). Many feminist and suffrage groups were Church-based and saw feminism as being their God-given calling (Tickner, 1987: 223) and some marched under the banner (Tickner, 1987: 72) of the Church League for Women's Suffrage, emblazoned with the embroidered slogan, 'the glorious liberty of the children of God'.

In contrast, the municipalised, positivistic, 'rational', secular urban planning that developed since the nineteenth century was more concerned with economic development and functionality than specifically with social issues or equality matters. So gradually religion, and women's issues too, dropped off the agenda of the emerging humanistic, material 'modern' society. Modern town planning, which gained stronger governmental and bureaucratic powers by the early twentieth century in the UK, has been characterised as a male-dominated, rational and scientifically based profession. As the twentieth century progressed, planning became driven by 'the white heat of technology' and concerns about the economy and the public realm (and motor car) (Greed, 1994). Policy became detached from the concerns of 'everyday life' which beset women, and gendered considerations were given a low priority.

Because of the planning profession's obsession with 'physical' land-use planning – as the only factor that really mattered – it has often been criticised for having little 'space' for so-called 'social issues' such as the differences between men and women's needs within the built environment, and little concern with 'cultural factors' such as the requirements of different religious and ethnic groups within society (Young and Stevenson, 2013). However, in the post-war era, there was a growing critique from a developing group of urban sociologists who argued that 'planning is for people' (Broady, 1968) (but whether 'people' included women is another matter (Greed, 1994)). But nowadays, at least within academia, there seems to be a move towards a softer, more reflective approach to understanding 'the city' and a mellowing of sociological thought to allow for diversity, equality, social inclusion and a wider range of social and cultural factors. Likewise town planning practice is gradually taking on board the importance of issues of, inter alia, diversity, equality, environmentalism, sustainability and community in determining policy making. One would think at first sight, this might lead to a greater recognition of both women's needs and the spiritual and religious aspects of urban society. But 'the problem' is that religion comes very low on the pecking order, as does gender, relative to other equality factors, such as sexuality, ethnicity and social deprivation (although arguably both gender and religion intersect with all these issues) (Reeves, 2005).

Many planners still retain some major 'blind spots' in their view of city and society (Young and Stevenson, 2013). Women's contribution to the economy and urban society is still 'beneath the radar' of many planners, but all the caring and home-making work undertaken still mainly by women produces and maintains the workforce in the first place (McDowell, 1983; Brownill, 2000). Likewise, 'religious capital' (Zohar and Marshall, 2004; Beaumont and Baker, 2011) – that is, all the work that religious organisations and their congregations undertake – actually 'fills in the gaps' when the declining Welfare State reduces services and provides a glue to keep society going. Faith communities provide essential social services, such as childcare, supplementary education, employment advice and help for the homeless and asylum seekers. This whole range of community facilities and activities contributes towards an area's regeneration and social development. In the aftermath of the Grenfell Tower fire in London, it was faith groups that first reached out. Religious buildings, ranging from Baptist churches to local mosques soon opened their doors to victims of all faiths and none, and acted as collection points for donations of food, clothing and other necessities. It is estimated that Christian churches alone in the UK provide over £3 billion of social work each year (Cinnamon Trust, 2015).

The challenge of sustainability

There are other challengers and challenges that will be mentioned in the course of the chapters. Planning policy has been taking on board the importance of 'sustainability', green issues and environmental matters. Curiously this can negate and detract from gender issues, and so many times in the recent past I have been told, 'we have done gender, you should be concerned with the environment' (Greed, 2005). But sustainability always had three components, social equity (including gender), economic vitality and environmental sustainability, so gender should always be a key component of any discussion about the environment and sustainability (Greed, 2016a). While this book is mainly about the urban situation, there has been a growing awareness of the fact that the environment, and environmental policy, are not gender-neutral, but that a gender lens must be applied to debates regarding environmental sustainability too (Buckingham, 2000; MacGregor, 2017). Likewise, one increasingly finds material on 'religion and the environment' (Crocket and Robbins, 2012), and even on religiously inspired ethics in the built environment (Fewings, 2008). We are all part of the wider environment, which encompasses both 'man-made' cities and the 'natural' environment, and sometimes it is difficult to separate the two, as many policy issues and policies, such as on air pollution or global warming, affect the whole planet, not just specific urban or rural locations. But, 'religion, gender and environment' are seldom combined. In comparison, some early second wave feminist books included significant religious components as well as spatial utopian dimensions and natural environment elements (Daly, 1973; Hayden, 1976). But we are encouraged that within social and feminist geography, there have been a few enlightened forerunner books which combine a discussion of religion with rural spatial matters (Maddrell et al., 2015; Hopkins et al., 2012; and especially Mee, 2016) as well as some mainstream theological books that tackle spatial issues (Sheldrake, 2014; Inge, 2003; Gorringe, 2002).

Trying to make sense of it all

So this book sits uneasily in the middle of all these different academic, professional and religious realms, lacunae and controversies. We cannot solve or resolve all this in one book, but we do want to increase awareness, create new synergies and further integrate both gender and religion into urban studies. I myself am a town planner, a feminist, but also a Pentecostal (by conversion in my late teens), from a free thinker background. To make sense of my life, I got drawn into urban sociology and have also explored urban theology for many years. The key to retaining my sanity is to be able to juggle all these different balls at once, while retaining a somewhat detached, but inquisitive perspective.

In the course of the various chapters the contributors give their own perspectives on the relationship between gender, cities and religion, from various, and sometimes conflicting, angles. For example, some chapters are concerned with the implications of including religion and gender in the urban policy mix, at the city-wide spatial level, in relation to the distribution and location of land uses. Others are much more concerned with the more detailed district and neighbourhood level of interaction in relation to urban design and morphology. It will become clear that many contributors question the idea that there is one generic, non-gendered, 'image of the city' (Lynch, 1968). Men are seen to be more dominant in public squares and main thoroughfares, whereas women often have their own preferred route-ways and focal meeting points, which interestingly are often related to the refuge and facilities religious buildings give them within the city of man. Other contributors give less attention to the spatial component, but discuss the aspatial (social, political and spiritual) implications for women of different urban planning policies in respect of the economy, community development, social inclusion, environmental sustainability and so forth. Some chapters focus specifically on gender and women's issues, whereas others provide a more generic, but essential, contextualising perspective on religion and the city. Likewise some contributors 'start' from a strongly religious standpoint, and their beliefs may constitute a challenge to the more secular, atheistic reader, but may also contradict and challenge the beliefs expressed in other contributors' chapters. But, all the contributors, in their various ways, try to tease out the relationships between gender, religion and the city.

Contents

Part I, 'Background', comprises two chapters: this one, which introduces the themes and contents, followed by a contextualising chapter by Chris Baker on postsecularism. Chris Baker, who is head of research at the William Temple Foundation (a body concerned with relating religion to public policy), provides a foundational theological and theoretical perspective on the paradigm shift towards postsecular cities that, arguably, provides the space to incorporate the inclusion of women's concerns and feminist perspectives within the changing religious landscape (Beaumont and Baker, 2011).

Part II, 'Women, Cities and Christianity', comprises four chapters. The chapters relate mainly to Europe and North America, and cover four of the main historical branches, namely Catholic, Protestant and Orthodox, plus a chapter on the Pentecostal and Charismatic movements. To give statistical context, Roman Catholics comprise over half of the world's 2.8 billion Christians, at around 1.2 billion adherents. The various Protestant denominations make up at least 1 billion, plus 260 million Eastern Orthodox believers. There are around 600 million Pentecostals and Charismatics, but some of these also belong to existing churches too, so are counted twice. In the UK, Black Pentecostal churches are the fastest growing, especially in cities. But similar patterns of growth are found in relation to other religions in the UK too (Gale, 2008; Peach and Gale, 2003). Globally, it is estimated there are around 1.8 billion Muslims, nearly a billion Hindus and about 10 million Jewish people globally. It is estimated there are over 3 million Muslims, around 1 million Hindus, half a million Sikhs, 290,000 Jewish people, 150,000 Buddhists and 25,000 Jains, within the total UK population of 63 million people (data from www.pewresearch.org/; www.brierleyconsultancy.com).

Chapter 3 is written by Dr Inés Sánchez de Madariaga a practising Catholic and a professor of urban planning in Spain (Madariaga and Roberts, 2013). She seeks to provide an overview on the position of women in the Catholic Church from an historical and contemporary perspective. She highlights the influence of Catholicism on the development of town planning and feminism from medieval times (Pizan, 1405) through to industrial utopian communities (Hayden, 1976) and on to modern day attempts to create new 'communities of intention' which embody Benedictine principles (Dreher, 2017). Throughout, she defends her beliefs. As a feminist, she makes the case for a male-only priesthood and the importance of the Virgin Mary (Warner, 1994).

In Chapter 4, Professor Petra Kuppinger provides an account of how the relationship between gender, cities and religion is working out within the Protestant Northern European context (Gmelch and Kuppinger, 2017). She describes how Lutheran churches are developing relationships with newly arrived Muslim communities in Germany. Her chapter illustrates the realities and role of faith groups within an increasingly multi-ethnic and religiously diverse modern Europe.

In Chapter 5, Dr Anna Papadopoulou, a Greek Cypriot architect, writing with her PhD supervisor, Juliet Davies from Cardiff University, draws on her recent research on the place of women in the city of Limassol in Cyprus. They illustrate how religious influences, past and present, have shaped women's place within the city, with reference to the effects of its past Muslim heritage, and the continuing influence of the Greek Orthodox Church on women's lives and urban form.

In Chapter 6 I first seek to define Pentecostal Christianity. As a Pentecostal convert in my late teens, in the second part of this chapter I will explain how gender, my faith and my urban spatial origins have interacted in my life and have led me into a career in town planning. (In Chapter 15 I will look at the problems Pentecostal churches encounter when dealing with the UK planning system.)

Part III, 'Women, Cities and Islam', comprises three chapters. The idea that the Muslim city is likely to be negatively spatially divided between male/female, public/private, inside/outside realms is prevalent in Western minds. But many Christian Mediterranean cities too have been shaped by similar forces (Domingo and Lopez, 1995), as illustrated in Chapter 5. The authors of these chapters show the situation is more complex and layered, sometimes to women's advantage. Indeed modern 'land-use zoning' in Western town planning is laden with ancient religious taboos as to the separation of men from women (whose influence arguably contributed to modern land-use zoning that separates work from home) (Greed, 1994: 79–82)

In Chapter 7 Professor Carmen Hass-Klau, a specialist in transport planning and the needs of the pedestrian (Hass-Klau, 2014), provides a contextualising historical perspective, comparing and contrasting urban form, street layout and planning principles of Christian and Muslim cities with reference to Spanish and North African cities. Her emphasis is more upon generic, rather than specifically gendered, factors in this overview.

In Chapter 8, the work of Mokarram Abbas, on the position of women in the Palestinian city of Nablus, is presented. She found that women were uneasy in male-dominated public spaces, but that there was a network of female-controlled spaces in local neighbourhoods, and that some of the safest spaces for women and accompanying children were small public parks (*mutanazahat*) (Abbas and Van Heur, 2014). She questions whether Western 'women and planning' principles are applicable to the Muslim city and whether separation, rather than mixing, of women and men in public space is the better solution.

In Chapter 9, the research of Roja Tafaroji on Afghani immigrant woman living in New Zealand is set out, in respect of the position of Muslim immigrant women in Auckland. The women are freed from the spatial and social restrictions of Kabul, but have to carve out their own territory and community within what is a very multi-ethnic, cosmopolitan city, where 'anything goes'. But many Afghani women still feel restricted because they are being 'observed' by older Afghani immigrants, although the younger generation in particular are increasingly adopting the life-style of their new country.

Part IV, entitled 'Non-Religious Space and Women', comprises two chapters. In Chapter 10, Katie McClymont looks at issues of religion and spirituality beyond organised faith systems and designated places of worship within postsecular society (Aune, 2015; Aune et al. 2017) in which few people attend church but many yearn for personal spirituality (Davie, 1994; 2015). She discusses the concept of municipal spirituality (McClymont, 2015), regarding the need for the planners to protect and provide non-denominational urban spiritual spaces, for reflection and memorial. She considers the role of religiously affiliated spaces and buildings, such as church halls that are used for play groups, childcare and community events. But because they are mainly used by women, they are 'beneath the radar' of the planners. Chapter 11, also by Katie, discusses the spatial, planning and gender implications of new age spirituality, paganism and environmentalism and the position of women therein. Such movements do not necessarily need dedicated spaces or buildings, but are often strongly linked to respect for nature and the countryside.

In Part V, entitled 'Planning Policy and Change', four contributors discuss the relationship of religion to the city and planning policy and practice. In Chapter 12, Andrea Mulkeen, an Anglican, who works for the property department of the Church Commissioners, reviews the extent to which the post-war UK planning system, from 1945 onwards, has taken religious land uses and activities into account (Mulkeen, 2014). She identifies the problems that the current planning system presents for those who wish to develop, expand or change the use of religious buildings, with reference to examples and suggestions for improving the situation.

In Chapter 13 Gill Reeve, drawing on her research and role as a night-church pastor, looks at the position of women and religion within the night-time economy and the 24-hour city. She discusses the role of the Church and other faith-based interventions in dealing with the problems encountered after dark by women. She highlights the problems she encountered as a woman minister setting up a new initiative to meet this gender-related need, in respect of the attitudes of both male-dominated established Church bodies and local authority 'urban governance networks'.

Chapter 14 is written by Carey Doyle, a researcher with CAG, a planning consultancy that has specialised in addressing the problems faith communities encounter dealing with local planning authorities (CAG, 2008). Her chapter examines the shortcomings of the planning system, not least the lack of clear central government guidance, and weaknesses in planning law, which have resulted in local churches and mosques being refused planning permission to build or change the use of buildings to accommodate their growing congregations.

In the final chapter (Chapter 15) I give a brief summary of the issues discussed. I offer a way forward, with reference to presenting a consciousness-raising 'toolkit' on how to mainstream religious issues into the planning system. The chapter is illustrated with reference to the 'KICC' mega church planning saga in London (Greed, 2016b). It sets out the problems encountered by such Black Pentecostal churches when dealing with the planners, not least because there is no statutory 'Use Class' specifically for religious buildings, no national survey of religious land-use needs, and the suspicious, unhelpful and secularist mentality held by many planners towards faith groups (Rogers and Gale, 2015)

References

Abbas, Mokarram and Van Heur, Bas (2014) 'Thinking Arab women's spatiality: the case of "mutanazahat" in Nablus, Palestine', *Gender Place and Culture* 21(10): 1214–1229.

Arthur, C. (1991) 'Diversion around the God slot: how can the humanities ignore the human soul?', *The Times Higher*, 8 November.

Aune, Kristen (2015) 'Feminist spirituality as lived religion: how UK feminists forge religio-spiritual lives', *Gender and Society* 29(1): 122–145. doi: 10.1177/0891243214545681.

Aune, K., Lovheim, M., Giorgi, A., Toldy, T. and Utriainen, T. (2017) 'Introduction: is secularism bad for women? La laïcité nuit-elle aux femmes?', *Social Compass* 64(4): 449–480. https://dx.doi.org/10.1177/0037768617727464.

Baines, B. (2009) 'Must feminists identify as secular citizens? Lessons from Ontario' in L. McCain and J. Grossman (eds) *Gender Equality: Dimensions of Women's Equal Citizenship*, Cambridge: Cambridge University Press, 83–106.

Beaumont, Justin and Baker, Chris (eds) (2011) *Postsecular Cities: Space, Theory and Practice*, London: Continuum.

Boot, Joseph (2014) *The Mission of God: A Manifesto of Hope*, Toronto: Freedom Press; St Catharine's, Ontario: Ezra Institute.

Broady, M. (1968) *Planning for People*, London: NCSS/Bedford Square Press.

Brownill, Sue (2000) "Regenderation: women and urban policy in Britain" in Jane Darke, Sue Ledwith and Roberta Woods (eds) *Women and the City: Visibility and Voice in Urban Space*, London: Palgrave, 114–129.

Buckingham, Susan (2000) *Gender and Environment*, London: Routledge.

Burgess, Stanley M. and van der Maas, Eduard M. (eds) (2002) *The New International Dictionary of Pentecostal and Charismatic Movements*, Grand Rapids, MI: Zondervan.

CAG (2008) *Responding to the Needs of Faith Communities: Places of Worship: Final Report*, London: CAG, Cooperative Advisory Group Planning Consultants.

Cinnamon Network (2015) *Cinnamon Faith Action Audit Report*, London: Cinnamon Network. www.cinnamonnetwork.co.uk/.

Cornwell, M. (2019) 'The rise of the secular', *Christianity*, May, 44–50. www.premierchristianity.com.

Cox, Harvey (1965) *The Secular City: Secularisation and Urbanisation in Theological Retrospect*, Harmondsworth: Penguin.

Cox, Harvey (1995) *Fire from Heaven: The Rise of Pentecostal Spirituality and the Reshaping of Religion in the Twenty First Century*, Cambridge, MA: Da Capo Press.

Crocket, C. and Robbins, J. (2012) *Religion, Politics and the Earth: The New Materialism*, London: Palgrave Macmillan.

Daly, Mary (1973) *Beyond God the Father: Towards a Philosophy of Women's Liberation*, Boston, MA: Beacon Press.

Davie, Grace (1994) *Religion in Britain since 1945: Believing without Belonging*, London: Blackwell.

Davie, Grace (2015) *Religion in Britain: A Persistent Paradox*, London: Wiley.

Domingo, Pilar Ballarin and Lopez, Candida Martinez (eds) (1995) *Del Patio a la Plaza: las mujeres en las sociedades mediterraneas* [From patio to town square, women in Mediterranean societies], Granada: Universidad de Granada .

Dreher, Rod (2017) *The Benedict Option: A Strategy for Christians in a Post-Christian Nation*, New York: Sentinel.

Eade, John and Mele, Christopher (eds) (2002) *Understanding the City: Contemporary and Future Perspectives*, London: Blackwell.

Fewings, P. (2008) *Ethics for the Built Environment*, London: Taylor & Francis.

Gale, R. (2008) 'Locating religion in urban planning: beyond race and ethnicity', *Planning Practice and Research* 23(1): 19–39.

Garbin, David and Strhan, Anna (eds) (2017) *Religion and the Global City*, London: Bloomsbury.

Gmelch, George and Kuppinger, Petra (eds) (2017) *Urban Life: Readings in the Anthropology of the City*, Long Grove, IL: Waveland Press (6th edition).

Gorringe, T.J. (2002) A *Theology of the Built Environment: Justice, Empowerment and Redemption*, Cambridge: Cambridge University Press.

Greed, Clara (1991) *Surveying Sisters: Women in a Traditional Male Profession*, London: Routledge, Chapter 2.

Greed, Clara (1994) *Women and Planning: Creating Gendered Realities*, London: Routledge.

Greed, Clara (2005) 'Overcoming the factors inhibiting the mainstreaming of gender into spatial planning policy in the United Kingdom', *Urban Studies* 42(4): 1–31.

Greed, Clara (2011) 'A feminist critique of the post secular city: god and gender' in J. Beaumont and C. Baker (eds) *Post-Secular Cities: Space, Theory and Practice*, London: Continuum, 104–119.

Greed, Clara (2016a) 'Religion and sustainable urban planning: if you can't count it, or won't count it, it doesn't count', *Sustainable Development* 24: 154–162.

Greed, Clara (2016b) *Christianity and Planning: Pentecostal Churches*, Occasional Paper 1, University of the West of England, Bristol.

Greed, Clara (2017) 'Sacred spaces' in S. Hawthorne (ed.) *Gender: God*, New York: Macmillan, 315–333.

Greed, Clara and Johnson, D. (2014) *Planning in the United Kingdom: An Introduction*, London: Palgrave Macmillan.

Habermas, J. and Ratzinger, J. (2007) *The Dialectics of Secularisation: On Reason and Religion*, Rome: Ignatius Press.

Harvey, D. (1975) *Social Justice and the City*, London: Arnold.

Hass-Klau, Carmen (2014) *The Pedestrian and the City*, London: Routledge.

Hayden, D. (1976) *Seven American Utopias: The Architecture of Communitarian Socialism, 1790–1975*, London: MIT Press.

Hopkins, Peter, Kong, Lily and Olson, Elisabeth (eds) (2012) *Religion and Place: Landscape, Politics and Piety*, London: Springer.

Inge, J. (2003) A *Christian Theology of Place*, Farnham: Ashgate.

Jaschok, Maria (2015) 'Gender and religion: gender-critical turns and other turns in post-religious and post-secular feminisms' in Anne Coles, Leslie Gray and Janet Momsen (eds) *The Routledge Handbook of Gender and Development*, London: Routledge, 47–57.

Kraemer, C.H. (2012) 'Gender and sexuality in contemporary paganism', *Religion Compass* 6: 390–401. doi: 10.1111/j.1749-8171.2012.00367.x.

Laurence, T. (ed.) (2014) *Good News for the Public Square: A Biblical Framework for Christian Engagement*, London: Lawyers Christian Fellowship.

Lynch, Kevin (1968) *The Image of the City*, Cambridge: MA and London: MIT Press.

MacGregor, Sherilyn (ed.) (2017) *Routledge Handbook of Gender and Environment*, London: Routledge.

Madariaga, Inés Sánchez de and Roberts, Marion (eds) (2013) *Fair Shared Cities*, London: Ashgate.

Maddrell, Avril, Terry, Alan and Gale, Tim (2015) *Sacred Mobilities: Journeys of Belief and Believing*, Farnham: Ashgate.

McClymont, Katie 'Postsecular planning: the idea of municipal spirituality', *Planning, Theory and Practice* 16(4): 535–554.

McDowell, L. (1983) 'Towards an understanding of the gender division of urban space', *Environment and Planning D* 1: 59–72.

Mee, Wendy (2016) 'The social lives of gender and religion: implications for development policy', *Sustainable Development* 24: 163–171.

Mulkeen, Andrea (2014) *The Planning System and Places of Worship: The Extent to Which the Planning System since the Second World War Has Facilitated the Needs of Faith Communities for Places of Worship*, London: University of Westminster.

Narayanan, Yamini (2016) *Religion, Heritage and the Sustainable City: Hinduism and Urbanisation in Jaipur*, London: Routledge.

Peach, C. and Gale, R. (2003) 'Muslims, Hindus, and Sikhs in the new religious landscape of England', *Geographical Review* 93: 469–490.

Pizan, Christine de (1405) *The Book of the City of Ladies* (1983), New York: Persea Books, translated by Earl Jeffrey Richards.

Reeves, D. (2005) *Planning for Diversity: Planning and Policy in a World of Difference*. London: Routledge.

Rogers, A. and Gale, R. (2015) *Faith Groups and the Planning System (England): Policy Briefing to Parliament*, London: Roehampton University.

Sheldrake, P. (2014) *The Spiritual City: Theology, Spirituality and the Urban*, London: Wiley.

Tickner, L. (1987) *The Spectacle of Women: Imagery of the Suffrage Campaign 1907–14*. London: Chatto and Windus.

Warner, M. (1994) *Alone of All Her Sex: The Myth and Cult of the Virgin Mary*, London: Picador Books.

Young, G. and Stevenson, D. (eds) (2013) *The Ashgate Research Companion to Planning and Culture*, Farnham: Ashgate.

Zohar, D. and Marshall, I. (2004) *Spiritual Capital: Wealth We Can Live By*, London: Bloomsbury.

2 The postsecular city

Site of emancipation or a false dawn?

Chris Baker

Abstract

This chapter critically explores the extent to which the idea of the postsecular, and in particular the postsecular city, has opened up a more creative discursive and empirical space by which to reflect on the role and experience of religious women in producing and planning the urban. After highlighting a genealogy of the concept of the postsecular city, the chapter goes on to investigate the emerging theme of postsecular feminisms, as a corollary to the postsecular city, before outlining some contrasting examples of postsecular urban feminisms by which to illustrate some of the breadth and complexity of this growing agenda. Examples of post secular feminist phenomena are given from within both the Christian and Muslim contexts, with reference, for example, to feminists challenging the relationship between the state and the Russian Orthodox Church, the role of the Virgin Mary as a radical figure, and also to the position of women within Muslim cities. This chapter provides a conceptual introduction and urban context to many of the issues discussed in this book.

Key words: feminism, postsecularity, intersectionality, agency, Islam, Russian Orthodoxy, cities.

Postsecular cities: the background and concept

The concept of the postsecular city has undergone a steady mutation since the 2011 volume of the same name that I edited with Justin Beaumont (Beaumont and Baker, 2011). Its publication marked roughly ten years since the 'postsecular' emerged as a term within Western political philosophy under the influence of Jürgen Habermas (2002; 2003). In the nearly 20 years since then, the use and conceptual framing of the term has mushroomed exponentially, and not uncontroversially, having moved beyond the narrow confines of Western political philosophy to become a term that has been critically discussed across most of the arts and humanities disciplines (Baker and Dinham, 2017), and indeed beyond the confines of the West (Cloke et al., 2019; Ratti, 2010). This globalised and thoroughly interdisciplinary discourse now includes vigorous and productive debate within both urban and feminist studies. This chapter explores the more recent developments of the use of the postsecular and postsecularity in these disciplines, as a way of setting some of the debates and themes covered in this volume within a broader context.

Postsecular cities: the theory and practice

When Justin Beaumont and I conceived of the *Postsecular Cities* volume in 2009, the conjoined use of these two words was still in its relative infancy. Up to this point, the terrain mapped by the concept of the postsecular had primarily engaged with the fields of political philosophy and sociology of religion. Here, its critical power centred on the unexpected re-emergence of religion and belief as powerful shapers of the public sphere in the late twentieth and twenty-first centuries. The postsecular appeared to call into question long-held assumptions regarding the teleological and linear logic of secularisation and in particular, the secularisation thesis around which sociology of religion had grouped most of its methodological assumptions. There is not the space, or indeed necessity, to rehearse the genealogy of the postsecular concept during this time, as this has been already done by several eminent commentators (Beckford, 2012; Calhoun et al., 2011; Taylor, 2007). But the oft-quoted sentence that seemed to crystallise the new intellectual and empirical terrain it signalled was Habermas' reflection in 2005 that it was now time for the West to 're-imagine a postsecular self-understanding of society as a whole in which the vigorous continuation of religion in a continually secularising environment must be reckoned with' (Habermas, 2005: 26).

The catalyst for Habermas' bold and reflexive re-imagining of his own classic Marxist, secular and liberal corpus (which had conspicuously ignored the role and impact of religion throughout most of the twentieth century) has both empirical and ethical origins. First is the resurgence of religion and belief as influential political global actors, especially those categories of religion that can be labelled as fundamentalist. Religion remains, Habermas observes, a culturally self-confident phenomenon that 'unmistakeably shapes the physiognomy [different characteristics] of the major civilisations' (2006: 1). But beyond this empirical and sociological fact, there is a moral and ethical dimension to the postsecular, at least in the West, which relates to issues of motivation and authority/legitimation. Following the untrammelled impact of what Habermas calls 'capitalist modernism', the Enlightenment virtues of rationality, autonomy and the right to private property (as a utilitarian means of redistribution of wealth) have become distorted and decoupled from the moorings of ethical and moral thought. The attendant stress on individualism as a social and cultural norm 'hollows out' the basis of solidarity and mutuality upon which fundamental concepts of justice depend. Religion, Habermas suggests, represents an indispensable source of moral and ethical wellsprings upon which the concepts of equality and liberty within a liberal democracy depend. Without this depth of commitment and vision that emerges at an ontological and therefore pre-political level of commitment, the liberal democratic state struggles to appeal beyond the materialistic present in its engagement with its citizens. This is because, in Habermas' view, the claims for its own authority are symbiotically entwined with the self-generating worldviews of religious traditions which in turn reflect the collectively binding ethical conventions that are necessary for the effective and democratic functioning of civil society (Habermas, 2010; 2011).

The philosophical and empirical roots of the postsecular had their most immediate effect on the religious field via the impact and role of faith-based organisations (FBOs) in the public square. The notion of the postsecular encouraged the exploration of how FBOs were developing new partnerships with secular actors, as well as generating new practices and technologies of care that became increasingly integral to a mixed economy of local service delivery (Baker and Skinner, 2006; Beaumont and Cloke, 2012; Dinham et al., 2009; Furbey et al., 2006; Jawad, 2012). These volumes and many others written at this time were instrumental in inserting the postsecular debate within critical, urban and human geography discourses, as well as social policy.

It was at this juncture that the *Postsecular Cities* volume was curated, and it sought to expand the then current debate in three ways. First, the production and reproduction of urban space within a globalised and diverse world was giving important clues as to how co-existing religious and secular impulses and worldviews (inherent in Habermas' earlier definition) were being practised in the real world. The distribution of power and knowledge intrinsic to these practices needed to be analysed in more nuanced and complex ways than early attempts describing the postsecular were able to grasp. Second, we wanted to create a body of work in which both the empirical and the theoretical were being brought into dialogue with each other in one place. The volume's global case studies of how the new co-existence between the religious, the spiritual and the secular is producing different urban publics were juxtaposed with theoretical voices attempting to articulate the new modernities and changes through which we are living. Third, we wanted to create a genuine and critical interdisciplinarity around the postsecular, which had to include theological alongside sociological, geographical, political philosophical and policy perspectives. Thus the volume had a four-fold structure entitled 'Mapping the Theoretical Terrain', 'Competing Experiences of Postsecular Cities', 'Postsecular Policies and Praxis' and 'Theological and Secular Interpretations'. In these three respects – the locating of the concept of the postsecular in the spatial and the material, the synthetic use of empirical case studies as well as theoretical frameworks, and its integral interdisciplinarity (including theology) – *Postsecular Cities* helped pave the way for the current sets of discourses and debates in this volume.

A typology of the postsecular

The current features of the current postsecular and postsecularity landscape have been expertly delineated by Elaine Graham (2019: 227–229). Her typology of the meanings of the postsecular relate to the different ways that the prefix 'post' has developed over the past 20 years.

The first use of the 'post' prefix is as in 'after'. This set of discourses critically assesses the impact of the re-emergence of religion and belief into public life including: the new post-metaphysical grounds for functioning democracies; the fragmentation of the nation state under the pressures of globalisation and new spaces for religion to emerge; and the retreat of the welfare state and the growth of new forms of voluntarism which favour faith-based organisations.

The second use is as in 'against', which refers to the critical revision of historical genealogies of the secular (initiated by Tala Asad and Jose Casanova in the early 1990s) which highlight the clear contextuality and cultural specificity of the Western Enlightenment and the secularism that emerged from it. Secularism's status as a normative, global and universal displacement of Christianity is now critiqued by voices from the Global South for the ideologies of violence and political domination it perpetrated in the name of progress and modernisation. Graham isolates some of the key motifs associated with this use of the 'post' prefix. 'Whose project was "the secular"? At what cost does it come? Deconstruction of categories and hermeneutics of suspicion towards grand narratives' (2019: 227).

This 'against' approach to the postsecular characterises much of the stance associated with postsecular feminisms. For example, Nandini Deo (2018) highlights new frameworks for understanding issues of agency, subjectivity and equality for religious women from the Global South who see themselves as both unambiguously religious and feminist in order to move beyond the 'imperial secularity' associated with colonialisation.

The third element of Graham's typology is that of 'beyond', which refers to the phenomenological turn in postsecular studies; i.e. the turn away from the abstract and reified readings of the religious and secular towards ways in which the identities and experiences associated with them are lived, embedded and performed in the public space. The close juxtaposing of multiple identities and epistemologies wrought by globalisation is creating new forms and expressions of the religious and the secular, some of which converge into interesting blurrings and agonistic or purposeful convergences. It is within these public spaces that the 'habitus' of the postsecular is generated.

Claiming notions of agency and political subjectivity in the public sphere

I am suggesting that the most productive areas of engagement for feminist and womanist readings of religion, belief and the urban are those that are reflected in the 'against' and 'beyond' dimensions of Graham's typology of the postsecular. But before outlining some exemplary case studies where these discourses are being played out, it is important to foreground the central questions this volume seeks to address. Namely, does the postsecular city as conceptual and empirical framework advance the recognition of the experience, knowledge and subjectivities of religious and spiritual women who also identify as feminists and as shapers of the public and urban realm? And, does it represent a genuinely new cultural, spiritual and political space for agency and subjectivity, or does it merely offer a thin veneer of these attributes?

There are understandably conflicting views on these questions. Clara Greed for example, does detect a more open and permissive discursiveness within certain academic discourses, as a result of the 'postsecular turn'. She discerns 'the opportunity to discuss and expose the dualism and disparities that have so hobbled women's lives, from a sociological, spatial and spiritual perspective … [and an] intellectual space [which] has opened up for mainstream researchers to ask value-laden and "subjective", even religious, questions as to whether the city is "good" and if so "for whom"' (Greed, 2011: 109).

However, for Greed, it is the prevailing situation on the ground that generally negates any opening up of the intellectual space, due to the entrenched patriarchalism and secularism of the planning profession. At the heart of this resistance to change is its inability to wean itself off the gendered private/public dichotomy that secular planning inherits from its Judeo-Christian roots. Public space is still inextricably linked to understandings of male agency and activity, while female experience is contained (or zoned) within either special public areas or within the domestic arena. The desire to regulate female sexuality and reproductive power is rooted in the need of a male self-image predicated on control and rationality (Greed, 2011: 109).

Alongside this inherited ideological perspective is the simple fact that most professional individuals responsible for planning the city tend to have a limited life experience and so are out of touch with 'the realities of everyday life that women, working class people, people with disabilities, the elderly and ethnic minorities experience as they battle their way through the modern city' (Greed, 2011: 109). Although women reflect the majority of 'the public' for whom planning is intended, the 'normal' citizen around whom planning needs are calculated is male.

> Thus, women tend to trip-chain their journeys as part of the multi-tasking they do every day to get everything done. If women's needs as workers and commuters were taken as seriously as men's in the planning policy-making process, this would have immense policy implications in reshaping citywide spatial policy.
>
> (Greed, 2011: 110)

Greed is also sceptical that debates around the 'postsecular' have specifically made any qualitative difference for women of faith and belief. She focuses on the role of Black Pentecostal women ministers and church leaders as an indicative cohort of religious women who 'apply their belief in the miraculous to the political situation' (Greed, 2011: 112). Their ability and commitment to span traditional public and private zonings mean these women exercise an ineluctable agency and political subjectivity as they become involved in community politics and local issues on the ground at a leadership level, often around equality and inclusion. However, this agency is often doubly invisible not only to central and local government, but also to the religious communities themselves whose patriarchal worldviews and culture deliberately or incidentally occlude the role and presence of women in public life.

For Elaine Graham, the potential root of lingering oppression for women of religion and belief within the postsecular lies ultimately more in its inability to think beyond the gendered nature of modernity. Its elevation of public reason involving ideas associated with 'impartial and non-contingent subjectivity and models of the free, self-actualising autonomous agent facilitated by the formation of liberal democracy' (2013: 54) are highly gendered and based on binary and essentialised representations of the male and female. Women, Graham reflects, 'may be the guardians of the world of affect and sensuality – along with that of reproduction – but the advancement of reason is a male task' (2013: 56). This legacy lives on in the 'Enlightenment feminism' that was associated with the second wave of feminism, which saw religion, particularly in its institutional forms, as a primary source of control of women's agency and political subjectivity.

Of course, as Graham points out, there has always been a heterogeneous tradition within feminist discourse that includes and embraces ideas and identities of religious belief and spirituality as intrinsic rather than extrinsic to women's experience (for example in the work of Luce Irigaray, Julia Kristeva, Donna Haraway and Grace Jantzen). Yet she maintains that the default position of feminist studies is to relegate religion and religious agency to those fields of discourse which can only enquire about them from the perspective of 'difference' or in respect of contexts that have yet to 'modernise'. These perspectives, Graham suggests, inhibit 'new explorations of how globalisation affects feminism as a political project and as a movement which proclaims and upholds human dignity and freedom by insisting that religion is ... an enemy of autonomy, authentic identity and progress' (2013: 59). Graham thus considers the postsecular to be insufficiently separate from its own Habermasian roots to act as a critically countervailing force that opens up new spaces for dialogue and genuine exchange. There is every danger that it becomes a concept that is simply squeezed into acting as an ineffective proxy term for ongoing secularism that is caught between 'the irresistible force of secularism and the immoveable object of religion, especially religious fundamentalism' (2013: 59).

A more positive reading of the liberative potential of postsecular feminism comes from Rosi Braidotti, who, perhaps because she is not starting from an overt standpoint of confessional belief, sees the landscape between feminism, religion, spirituality and the secular in more fluid and open-ended terms. In her genealogical excavations she traces the spiritual not only within feminist writings since the earliest pioneers, but also within critical theory itself, which inherits its teleological imaginaries of social progress and future evolution from its own Judeo-Christian roots (Braidotti, 2008: 9). Other spiritual features within critical theory, derived primarily from the psycho-analytical tradition, include the vitality of deep human drivers, the significance of symbolic or totemic figures as sources of internal and social cohesion, together with the importance of emotional layering in the process of subject formation. This layering includes 'the affective, unconscious and visceral elements of our allegedly rational and discursive belief system' (Braidotti, 2008: 11).

With these spiritual dynamics already embedded within critical theory, Braidotti senses that the postsecular is a new and discursive framework that encourages a positive focus on resistance, counteraction or transcendence. This new perspective, harnessing the latent spirituality within critical theory, captures a neo-vitalist worldview that creates the urge for seeking out radical relationality as a form of affirmative empowerment, defined by Braidotti as 'the ethical ideal … to increase one's ability to enter into modes of relation with multiple others' (2008: 16). Within this framework, Braidotti proposes an expression of agency and political subjectivity that 'consists of multiple micro-political practices of daily activism or interventions in and on the world we inhabit for ourselves and future generations' (2008: 16). It is this wider reappraising of the spiritual genealogies of critical theory, and its subsequent ethical impulses to imagine and endorse new expressions of agency and cross-silo relationships, that creates the optimum conditions for postsecular feminism. Under these conditions, performative ethics, imaginary and experience can be actively nurtured and affirmed, and 'the question of faith in possible futures' (2008: 18) can be validated and discerned.

Urban postsecular feminisms

In this section I bring the foundational debates outlined above into an overview of some of the different ways the agency and subjectivity of religious women in urban public spaces are being theorised and researched. This is not intended to be a representative list. This is, after all, an area of enquiry still in its infancy. However, this final section highlights some of the diverse ways the phenomenological and material practices of religious women in urban settings are now becoming more visible and valorised.

One simple fact worth iterating at this point, since it is now a foundational truism that supports the whole trajectory of this volume, is that we are now aware of how deeply the 'religious' produces the 'urban', just as the 'urban' produces the 'religious'. This insight informs an innovative and interdisciplinary volume entitled *Global Prayers: Contemporary Manifestations of the Religious in the City* which seeks to explore, through multi-media representations, how 'New religious practices permeate public spaces and the cultural scenes of these metropolises. Religious communities inscribe themselves into urban life acoustically, typographically, and architecturally. They generate their own aesthetics and forms of consumption' (Becker et al., 2013: back cover). Within this methodological and epistemological framework, urban anthropologist Birgit Meyer reflects on how Habermas' category of the postsecular 'acknowledges the presence, force and potentially positive influence of religion in the public domain' (p. 595). This recognition of religion as a public force requires a new critical framework that studies how religion and belief are historically and culturally situated phenomena, and how 'religion' and the 'city' are 'interrelated in various settings'. According to Meyer, this newly-understood symbiosis generates two major research questions: 'How do new religions transform urban space?' and on the obverse, how do 'Cities generate specific forms of religion?' (p. 595).

This volume will critically engage these two questions as it maps and analyses how the embedded and embodied practices of religious women, and indeed, the lived-out beliefs of all women, produce the urban. It is in this spirit of enquiry that I propose three brief case studies.

Public piety within secularised spaces

The first case study in Becker et al. (2013) engages with the symbolic and cultural power associated with the public wearing of the veil by Muslim women, especially young women for whom it appears a more conscious and politicised act. This agenda, established by Muslim scholars such as Saba Mahmood (2005), reflects the re-emergence of piety as a powerful counter-symbolic understanding of agency and subjectivity. Of course, there is much more to a reading of Muslim cities and the place of women within them than the issue of the veil, as discussed in later chapters of the current volume by Carmen Hass-Klau and Roja Tafaroji. These combined perspectives challenge the hegemonic perceptions of secularism and feminism, both of which are interpreted as contingent dimensions of Occidental colonialism rather than universal norms. Mahmood's theoretical framework emerges from research mapping the public engagement of Muslim women in the Islamic revivalism associated with the Egyptian mosque movement in the mid-1990s. This period saw the rapid proliferation of neighbourhood mosques, which as well as offering new sources of welfare, medical aid and education at the local community level, also provided opportunities for woman-to-woman education in respect of understanding core Qur'anic principles and how they might be applied on a day-to-day (or mundane) basis.

Mahmood argues that this embracing of piety across class and educational divides furnishes a wider reading of feminist empowerment than that on offer within Western feminism – what she refers to as 'multiple modalities of agency' (2005: 154). Her ethnographic research 'attends to the specific logic of the discourse of piety that is not located in the intentionality of actors, but in the relationships that are articulated between word, concepts and practices of a discursive tradition' (2005: 16). Within this more embedded and connected framework of tradition, Islamic concepts that shape bodily and spiritual self-realisation such as *al-haya* (referring to shyness, modesty and diffidence) become positive expressions of ethical formation that contribute towards the reaching of one's true potential (2005: 155). In Islamic ethics, the Western disjunction that is assumed to exist between desire and social norms is often collapsed. Desire and social norms need not be uncoupled but can be mutually affirmative. As Mahmood says, 'the distinction between the subject's own desires and socially prescribed performances cannot be easily presumed' and 'submission to certain forms of external authority is a condition for achieving the subject's potentiality' (2005: 31).

Mahmood's work has been critiqued from both secular and postcolonial strands of feminism. Rosa Vasiliki suggests that the embedded notions of agency that Mahmood identifies, rooted within the habitus of religion and culture, ultimately reinforce the status quo and so produce apolitical and non-emancipatory outcomes for non-Western women. It fails to engage adequately with 'questions of inequalities and the ideologies that sustain their natural status' (2016: 119). Afiya Shehrbano Zia (2018) meanwhile claims that postsecularist feminists are abandoning the category of class far too readily in their analysis of the public square. Zia suggests that Mahmood's theorising emerges from the comfortable positionality associated with the diasporic cosmopolitan (2018: 161), whose work and experience no longer engages with the palpable danger that secular women face, for example, in promoting healthcare reforms such as polio vaccination, in authoritarian theocratic societies.

However, Mahmood is under no illusions as to the pressures that living out of pious identities creates for religious women subjects, even in the overall context of an Islamic revival. It is by no means a smooth or untroubled transaction. Muslim women related to her the pressures they felt as working women in a secular environment, negotiating for example interactions with unrelated male colleagues, using public transport in mixed compartments, overhearing conversations that are impious or coarse, as well as experiencing strong reactions from other (often male) family members who were opposed to more stringent expressions of religious devotion (2005: 155).

Some of the ambiguous complexity associated with the public performativity of piety in other societies is picked up by secular anthropologists and sociologists (including Gökariksel and Secor, 2015; Navaro-Yashin, 2002). A detailed example of this is the work of Nilüfer Göle and her reading of how the secular imaginary shapes the identity of Turkish society. At a bureaucratic level, Turkey operates under a French imported system of *laiklik* or laïcité which was promulgated after the collapse of the Ottoman Empire by Kemal Ataturk in the 1920s, and which forbids the involvement of the state in government affairs and vice versa. However, the Islamic revival of the late 1970s has re-harnessed expressions of Turkish nationalism in the direction of a more neo-Ottoman, rather than secular, vision of Turkish global prowess.

According to Göle, some young Muslim women's aspiration to wear the headscarf in university campuses (for example) reflects the search for both a modern and an Islamic identity. This search necessitates religious citizens undertaking a 'voluntary secularism' (2010: 258), which involves presenting a 'civilised self' within the phenomenology of a secular habitus that influences everyday life. This habitus is promulgated through the legal and education systems, and reinforced through the construction of secular urban spaces to rival the influence of religious ones, such as ballet houses, concert halls and ballrooms. It is considered to be a higher form of civic belonging. The aspiration of young Muslim women to access a secular higher education could be further evidence of this 'secularising' trajectory, implying a desire to leave the conformity and restricted intellectual access of the religious field.

However, this tactic of voluntary secularism should not be misconstrued as a simple form of covert ideological capitulation or dissonance. For Göle, the wearing of the headscarf represents a public performance of both feminine and religious visibility. This challenges not only the hegemony of the secular over definitions of self, sexuality and space, but it also challenges religious hegemonic assumptions by breaking the normative taboos and assumptions of women wishing to claim public agency in 'secular space'. Her conclusion is that these young Muslim women are deploying what she calls (after Victor Turner) 'performative reflexivity' by which 'perceptive members of a socio-cultural group turn back upon themselves, upon relations, actions, symbols, meanings and codes ... to make up their public selves' (2010: 263). This framework perhaps enables a more symbiotic and dialectical reading of women's religious power in 'secular' modernities, whereby Islam can be seen to be shaped by the secular age, but in turn produces alternative notions of the self, morality and piety that shape or rearrange the 'habitus' of the secular.

Postsecular feminism and oppositional postsecularity

If the previous case study explored the edgy and ambiguous power of notions of female piety to shape the urban sphere in the context of everyday life, then a Russian example highlights a more overtly political strategy of religious and feminist disruption of both religious and public (or rather state-controlled) space as a one-off and highly stylised postsecular event. In a case study that still reverberates several years after its enactment in 2012, a female punk collective called Pussy Riot entered the Russian Orthodox Cathedral of Christ the Saviour in Moscow to perform what they called a punk prayer entitled 'Mother of God, Banish Putin' on the raised platform in front of the doors of the iconostasis. The punk prayer included the following lyrics: 'Virgin Mary, Mother of God, banish Putin; Banish Putin, banish Putin! Congregations genuflect, black robes brag; Golden epaulettes, freedom's phantom's gone to heaven' before ending with the specific plea to the Virgin Mother to 'Become a feminist, we pray thee' (wsxokn, 2013).

In a recently published volume, I unpack both the intent and the interpretation of the event from a postsecularity lens (Cloke et al., 2019). Was the invocation to the Virgin Mary to banish Putin an attempt to generate blasphemy and offend the Russian Orthodox Church? Was it an ironic and postmodern form of political protest that lampooned political authoritarianism and religious patriarchy? Or was it in fact a genuine, if somewhat unconventional, liturgical act of prayer? A traditional approach along the lines of mainstream feminism would probably focus on the second of these three questions at the expense of the other two. A postsecularist approach I suggest would countenance the possibility of all three readings, and that the intentional, and the ongoing power and resonance of the event lies precisely in the way it consciously and skilfully blends all three perspectives, and perhaps many more besides. A press statement from the Pussy Riot collective, released after the arrest of three of the original protagonists, Nadezha Tolokonnikova, Maria Alyokhina and Ekaterina Samutsevich, stated:

> In all our public statements, we constantly emphasise that the punk prayer 'Mother of God, Banish Putin' was truly a prayer – a radical prayer directed to the Mother of God to prevail upon the earthly authorities and ecclesiastical authorities who take their cue from them. Among the two dozen Pussy Riot members, many are Orthodox believers for whom a church is a deep place of prayer … we did not desecrate the church nor did we blaspheme … We passionately prayed to the Mother of God, asking her to give us all the strength against our incredibly merciless and wicked overlords. And we will continue to sing songs and will pray for those who want us killed and thrown into prison, because Christ teaches us not to wish death or prison on those whom we do not understand.
>
> (quoted in Uzlaner, 2014: 28)

Then there is the question of *interpretation*. The self-identification of some Pussy Riot members as believers in the Orthodox Church created 'a new interstitial quagmire of theological, legal and constitutional judgements' (Cloke et al., 2019: 167). Indeed Maria Alyokhina, in her book 'Riot Days', which covers the events and her subsequent time in prison, frames the intention throughout it that their 'riot' was informed by both feminist and Christian principles, not least to free the Virgin Mary from state control and celebrate her as a feminist icon (Alyokhina, 2017). In an attempt to downplay the political efficacy of the event and prevent future copy-cat demonstrations, the Russian state's line of attack focused on the moral and spiritual outrage caused by the women to the sensibilities of a newly created monolithic imaginary of Russian Orthodox believers. It also indicted them for undermining 'the spiritual foundations of the state' (Cloke et al., 2019: 167). On this basis they were prosecuted for violating canon law, dating back to the seventh century (the Council of Trullo) for which they served two years in a minimum-security prison. Secular law on its own was paradoxically inadequate to secure prosecution, as the women would not have infringed public order or safety regulations if they had done the punk prayer outside the territory of the Orthodox Church.

Dimitry Uzlaner (2014), who has done extensive research on this case, suggests that postsecularisation in Russia reflects fresh interpenetrations of the religious and the secular, which are creating new hybrid spaces and social and political imaginaries. 'Pro-authority' hybridities are when religious and secular agencies combine to support and legitimate the authority of the state. By contrast, 'oppositional' hybridities are those assemblages of religious and secular social actors who create new spaces and processes which challenge the authority of the state and the basis for its claim on the undivided loyalty of its citizens and institutions.

Pussy Riot are therefore a case study of postsecular feminist insurgency that deploy religious tropes and rituals to inscribe alternative readings of agency and political subjectivity in the modern global city.

Muslim gated communities in the exurbs: new intersections of class, religion and gender

My final case, from *Global Prayers* (Becker et al., 2013), offers an example of the intersectionality of class, religion and urban planning, and centres on research undertaken by a self-identified secular housing activist Ayşe Çavdar. In it she sets out to explore what political Islam means to those now living in the new enclaves and gated communities that have sprung up around Istanbul to cater for the needs of a growing devout Turkish middle class, some of which are being built by the Islamic RP or Welfare Party.

Many citizens in these new enclaves can trace their origins to residents of the working-class shanty towns or *gecekondu* that were created on the edge of Turkey's major industrial cities. These towns were populated by workers from the rural hinterlands, brought in to build the physical infrastructure to support Kemal Ataturk's vision of a secular and cosmopolitan Turkey. It was from these areas that a newly politicised Islamic identity emerged that eventually became the political base for the current and longstanding President of Turkey, Recep Erdogan, whose self-proclaimed agenda is to bring Islamic principles into education as part of the creation of a pious generation (*Financial Times*, 2018)

Çavdar's research involved her setting up a women-only Qur'anic reading group so that she could engage in conversation with local women. There were very few public spaces or amenities in the area, and those that were there (for example children's play parks) were rarely used. Women who were left at home while their husbands went to work did not feel comfortable with being potentially observed in public places. After spending some time in Qur'anic study, Çavdar notes the hostility with which she is met by some of the group when she explains who she is and what she wants to discuss. Two members simply refused to talk to her. But eventually she managed to glean some information about their experience of these urban spaces, and the general view was not positive: 'The others [in the reading group] began to tell me of their experiences of Başakşehir: nobody knows each other, no one intends getting to know each other; public transportation is very bad' (2013: 205). However these negative social impacts appeared to be offset by other more positive features that suited this group very well: 'it was quiet, peaceful, planned, secure and a good place to raise children were among the reasons given' (2013: 205).

The gated community, although devoid of public facilities and diversity, in many ways offers the ideal environment that enables these women to find their place in ways in which Greed has not noticed within Western secular planning. Above all, Çavdar notes, it is the intersection of class and religion that makes these spaces resonate for these Muslim women.

> My interviews with the pious and middleclass of Başakşehir: [showed] it was obvious that they were not keen on the perception that they were poor and excluded from the city. They had come ... to enjoy their recently elevated position, won through collective political struggle and hard work, which had boosted the national economy while also improving the local economic situation. They did not want their lifestyle judged by the secular lobby or by those with traditional understandings of piety, who looked down on the conspicuous exhibiting of newly acquired social positions and power ... they wanted respect, not only for their religious and ideological backgrounds but also because of their success commercially. A woman explained to me how it is possible to merge religious and profane values into one value system. 'Allah en iyi nimetlerini ona inananların üzerlerinde görmek ister' ('God wants to see his best blessing in the believers').
>
> (Çavdar, 2013: 207)

In this case study we see an intersectionality of religion, class, gender and ideology that has a direct impact on the urban environment in the creation of a space that while sharing many features associated with exurban communities, also feels a safe enclave in which public and private expressions of piety and wealth can be seamlessly conjoined.

Conclusion

In this chapter, I have taken an historical overview of the genealogy of the concept of postsecular cities, tracing its early origins in political philosophy and sociology, where its arguments are primarily critical of Enlightenment paradigms of modernity and the social and economic imbalances that have been caused as a result. There is widespread recognition that these paradigms now work against the experiences and rights of several key public constituencies including religion, belief and gender. More recent developments include a wider disciplinary engagement (including feminist studies, cultural and human geography, theology, religious studies, policy studies etc.). But developing Graham's important typology, we can also see how the postsecular, via the category of postsecularity, moves beyond the primarily cognitive, into the material, the spatial, the affective and the ethical.

These moves, I would argue (along with Braidotti) create a more genial and open space for religious postsecular feminisms to emerge, including a sympathetic understanding of how religious women reproduce the urban and vice versa. Much work however, remains to be done creating a more level methodological and theoretical playing field vis-à-vis other strands of feminist and planning discourse. I have concluded this chapter with three brief case studies that exemplify new types of discourse across different disciplinary perspectives of how religious women, under the conditions of postsecularity, reproduce urban spatialities and configurations. They are representative of the growing diversity of experiences, perspectives and theoretical frameworks that are now beginning to emerge.

References

Alyokhina, M. (2017) *Riot Days*, London: Penguin.

Baker, C. and Dinham, A. (2017) 'New interdisciplinary spaces of religion and belief in contemporary thought and practice: an analysis', *Religions* 8(2): 16. doi: 10.3390/rel8020016.

Baker, C. and Skinner, H. (2006) *Faith in Action: The Dynamic Connection between Religious and Spiritual Capital*, Manchester: William Tempe Foundation.

Beaumont, J. and Baker, C. (eds) (2011) *Postsecular Cities: Space, Theory and Practice*, London: Continuum.

Beaumont, J. and Cloke, P. (eds) (2012) *Faith-based Organisations and Exclusion in European Cities*, Bristol: Policy Press.

Becker, J., Klingman, K., Lanz, S. and Wildner, K. (eds) (2013) *Global Prayers: Contemporary Manifestations of the Religious in the City*, Baden: Lars Müller Publishers.

Beckford, J.A. (2012) 'SSSR presidential address: public religions and the postsecular. Critical reflections', *Journal for the Scientific Study of Religion* 51(1): 1–19.

Braidotti, R. (2008) 'In spite of the times: the postsecular turn in feminism', *Theory, Culture and Society* 25(6): 1–24.

Calhoun, C., Jurgensmeyer, M. and van Antwerpen, J. (eds) (2011) *Rethinking Secularism*, Oxford: Oxford University Press.

Çavdar, A. (2013) 'Negotiations as a research methodology' in J. Becker, K. Klingan, S. Lanz and K. Wildner (eds) *Global Prayers: Contemporary Manifestations of the Religious in the City*, Baden, Lars Müller Publishers, 198–215.

Cloke, P., Baker. C., Sutherland, C. and Williams, A. (2019) 'Wider religious and spatial conditions' in *Geographies of Postsecularity: Re-envisioning Politics, Subjectivity and Ethics*, London and New York: Routledge, 155–184.

Deo, N. (ed.) (2018) *Postsecular Feminisms: Religion and Gender in Transnational Context*, London and New York: Bloomsbury.

Dinham, A., Furbey, R. and Lowndes, V. (eds) (2009) *Faith in the Public Realm: Controversies, Policies and Practices*, Bristol: Policy Press.

Financial Times (2018) 'Erdogan's "pious generation" goal drives Islam into education', 29 May. www.ft.com/content/83328a4e-4317-11e8-97ce-ea0c2bf34a0b.

Furbey, R., Dinham, A., Farnell, R., Finneron, D. and Wilkinson, G. (2006) *Faith as Social Capital: Connecting or Dividing?*, Bristol: Policy Press.

Gökariksel, B. and Secor, A. (2015) 'Post-secular geographies and the problem of pluralism: religion and everyday life in Istanbul, Turkey', *Political Geography* 46: 21–30.

Göle, N. (2010) 'The civilizational, spatial and sexual powers of the secular' in M. Warner, J. Vantantwerpen and C. Calhoun (eds) *Varieties of Secularism in a Secular Age*, Cambridge, MA: Harvard University Press, 243–264.

Graham, E. (2013) *Between a Rock and a Hard Place: Public Theology in a Post-secular Age*, London: SCM Press.

Graham, E. (2019) 'Interrogating the postsecular' in J. Beaumont (ed.) *The Routledge Handbook of Postsecularity*, London and New York: Routledge, 223–232.

Greed, C. (2011) 'A feminist critique of the postsecular city: god and gender' in J. Beaumont and C. Baker (eds) *Postsecular Cities: Space Theory and Practice*, London and New York: Continuum, 104–119.

Habermas, J. (2002) *Religion and Rationality*. Edited by E. Mendieta. Cambridge: Polity Press.

Habermas, J. (2003) *The Future of Human Nature*, Cambridge: Polity Press.

Habermas, J. (2005) 'Equal treatment of cultures and the limits of postmodern liberalism', *Journal of Political Philosophy* 13(1): 1–28.

Habermas, J. (2006) 'Religion in the public sphere', *European Journal of Philosophy* 14(1): 1–25.

Habermas, J. (2010) 'An awareness of what is missing' in J. Habermas et al., *An Awareness of What Is Missing: Faith and Reason in a Post-secular Age*, Cambridge: Polity Press, 15–23.

Habermas, J. (2011) '"the political": the rational meaning of a questionable inheritance of political theology' in E. Mendieta and J. Van Antwerpen (eds) *The Power of Religion in the Public Sphere*, New York: Columbia University Press, 15–33.

Jawad, R. (2012) *From Wellbeing to Ways of Being: Religion and Faith-based Welfare in the UK*, Bristol: Policy Press.

Mahmood, S. (2005) *Politics of Piety: The Islamic Revival and Feminist Subject*, Princeton, NJ: Princeton University Press.

Meyer, B. (2013) 'Lessons in "global prayers": how religion takes place in the city' in J. Becker, K. Klingen, S. Lanz and K. Wildfner (eds) *Global Prayers: Contemporary Manifestations of the Religious in the City*, Zurich: Lars Muller Publishers, 591–599.

Navaro-Yashin, Y. (2002) *Faces of the State: Secularism and Public Life in Turkey*, Princeton, NJ: Princeton University Press

Ratti, M. (2013) *The Postsecular Imagination: Postcolonialism, Religion and Literature*, London and New York: Routledge.

Taylor, C. (2007) *A Secular Age*, Cambridge, MA: Harvard University Press.

Uzlaner, D. (2014) 'The Pussy Riot case and the peculiarities of Russian post-secularism', *State, Religion and Church* 1(1): 23–58.

Vasiliki, R. (2016) 'The politics of postsecular feminism', *Theory, Culture and Society* 33(2): 103–123.

wsxokn (website editor) (2013) Punk Prayer English translation lyrics. https://genius.com/Pussy-riot-punk-prayer-english-translation-lyrics (accessed 3 April 2019).

Zia, A. (2018) 'Why I am not a postsecular feminist: Pakistan, polio and the postsecular' in N. Deo (ed.) *Postsecular Feminisms: Religion and Gender in Transnational Context*, London and New York: Bloomsbury, 155–174.

Part II

Women, cities and Christianity

3 An overview of Catholic Christianity and women and space

Past and present

Inés Sánchez de Madariaga

Abstract

This chapter explores key issues at the intersection of the three topics of the book – religion, women and space – within Catholic Christianity, past and present. Because of the wide scope of the ground to be covered, both historically and geographically, this text cannot present a systematic or overarching panorama of women and space within Catholicism but provides an overview. The chapter is divided in three sections. The first provides some preliminary observations on women and space in the New Testament. The second section looks at the role of women in the Church, and addresses the issue of a male-only priesthood. The final section illustrates how women have contributed to building model community spaces of religious practice within Catholicism.

Key words: women, space, Catholic Church, leadership, community building, gender equality

Introduction

A great deal has been published on women, religion and space (including architecture and cities): as separate, relatively disconnected topics, but far less on the inter-relationships between these factors. For example, much has been written about church architecture and its meanings (McNamara, 2009). In the Middle Ages Christianity permeated Western culture, and city design was religiously inspired (Mumford, 1965). Scholarship on the history of daily life has uncovered many aspects of the social history of women in Christianity and the lives of prominent individual women (Duby, 1990; Martínez, 2000). There has been little research (such as Schulenburg, 2005) on how the three factors inter-relate within Catholicism.

This chapter focuses on the Roman Catholic Church and the position of women therein from a gender-aware perspective. This church is the longest-lived and the most organised of the three branches of Christianity, with a global reach and a 2,000-year history (Brennan, 2016). A gendered research approach usually looks at the representation and participation of women in the organisation under consideration and then at the level of recognition of the issues affecting women's lives therein and the ways of ameliorating the situation, for example in my work on women and urban planning (Sánchez de Madariaga and Roberts, 2013). This chapter will use this distinction as its main analytical approach addressing women's participation and representation. The first section looks at the depiction of women and space in the New Testament. The second section discusses the role of women in the Church, and sets out the case for a male-only priesthood. The last section illustrates how women have contributed to building community spaces of religious practice within Catholicism.

Women and the gospel

Women in the New Testament

The New Testament of the Bible narrates the story of Jesus of Nazareth, an itinerant preacher who lived 2,000 years ago in a distant province of the Roman Empire. A Jewish rabbi, he had no home or family of his own, and was followed by a group of women and men who had left their jobs, fathers, husbands and homes (Fee and Stuart, 1981).

References to women in this text abound. They appear to be very relevant actors in extraordinarily positive ways for the society of the time. After his birth, an elderly woman named Anna recognises the baby Jesus as the Messiah, and prophesies over him as does the better-known male prophet, Simeon (Luke Ch. 2, Ver. 25–40). During his lifetime, we see women leaving house and husband to follow Jesus, some of them from wealthy backgrounds who provided financial support (Luke Ch. 8, Ver. 1–3). Mary and Martha her sister often entertained him in their home (Luke Ch. 10, Ver. 38–42). But Jesus also associated with 'less acceptable' women such as the Samaritan woman, of dubious sexual reputation, who Jesus engaged in conversation (John Ch. 4, Ver. 4–42). He gave her spiritual teaching on a personal level, which, at the time, would normally only take place between men, and only between Jews at that. He welcomed a woman who anointed him with costly perfume (anointing was a sign of priesthood) and who kissed Jesus' feet (Mark Ch. 14, Ver. 3–9). We read about women at the margins, many of them outcasts and untouchables – widows (Luke Ch. 7, Ver. 11–17), prostitutes or adulterers (John Ch. 8, Ver. 1–11), chronically menstruating women and other 'unclean' women (Matthew Ch. 9, Ver. 20–22) being greeted by Jesus with affection. He also intervened when a woman 'taken in adultery' was to be stoned to death (John Ch. 8, Ver. 1–11). Jesus spoke of marriage as a covenant between a man and a woman as two equals made in God's image and likeness (Matthew Ch. 19, Ver. 1–12) as opposed to the Mosaic law which allowed men unilaterally to divorce women.

Women were the ones who did not fearfully flee when things got tough and dangerous, along the Via Crucis (The Way of the Cross) from the Mount of Olives up to Golgotha. Women are present at his Crucifixion, including Jesus' Mother Mary (John Ch. 19, Ver. 25). Women were the first ones there at the Resurrection, who bare testimony of the event (Matthew Ch. 28). Yet giving testimony in Antiquity was a male-only privilege. Mary Magdalene was the one who carried the good news of Jesus being alive, but the disciples did not believe the women until they saw for themselves.

After the Ascension of Jesus to Heaven, Mary the mother of Jesus heads the group of those women and men who receive the Holy Spirit at Pentecost (Acts Ch. 2). Later, women are leaders in the first Christian communities. The first Christian gatherings described by Luke in the Acts of the Apostles met in houses headed by women, such as that of Mary, mother of Mark, whose female servant Rhoda opens the door for Peter when he is freed from jail (Acts Ch. 12, Ver. 13–15). In the books and letters within the New Testament, which describe the first decades of expansion of the Christian faith within the Roman Empire, women often appear on equal ground with men among those receiving and transmitting the faith (Acts Ch. 12, Ver. 13–15), praying, prophesying, preaching, organising meetings and ministering to others, as well as being among the first martyrs of the persecutions. Women, together with slaves and foreigners, that is, those not in positions of public authority, are the first to convert in large numbers, but also some high status Roman Patrician women believe too (Acts Ch. 16, Ver. 14–15 and Acts Ch. 18, Ver. 2–3). These texts imply an astounding qualitative change in attitudes towards women, among Christians, compared with the standards of the time.

Women and the 'Kingdom'

As Christianity developed, believers tried to live in the Kingdom of Heaven on a daily basis ('Thy Kingdom Come on Earth as it is in Heaven'), but the relationship between women and men needed to be clarified. St Paul explained, there are no distinctions between women and men: all human beings are radically equal in the Kingdom, whether Jews or Gentiles, men or women, slaves or free persons (Galatians Ch. 3, Ver. 28). That did not mean women and men had the same roles in the Church; however, women did have a wider range of roles than was customary in society at that time.

As the centuries progressed, a large number of women were recognised as saints by the Church, many of whom are still popular subjects of devotion. The earliest saints are the martyrs who had suffered death because of their faith. The Congregation for the Cause of the Saints is the Vatican organisation charged with defining the procedures and setting the models for the different steps towards canonisation, that is who can become a saint. To grow in sanctity is to grow in communion with the Holy Spirit. Beyond the initial stages of salvation, sanctification is the primary aim of the Christian Catholic life, and sainthood (not priesthood) the most important state for Catholics. It is highly significant that no difference here exists between the sexes.

There were also many women who were recognised as Catholic mystics, those individuals who received glimpses during physical life of what this non-physical, eternal Kingdom might be like. Some of the greatest mystics have been women. Following the scriptural century-old tradition of using place as a metaphor for spiritual things, the great mystic Saint Theresa of Avila explains the inner experience of sanctification in architectural and spatial terms in her book of 1577 (Theresa, 2008). She writes of a crystal castle with seven chambers, its many rooms as a metaphor to describe her ineffable internal spiritual state.

Finally, and particularly important, the first human being to have fully reached 'The Kingdom' (Heaven) is a woman, namely the Virgin Mary, mother of Jesus (Luke Ch. 1, Ver. 46–55). The Virgin Mary precedes all humanity in reaching this place of peace and love, of eternal life and salvation. She is 'full of grace' through her humility, her faith is 'without ego'. She is free of 'sin' (sin is defined as *hamartia*, that is, missing the mark in the original Greek). She is the 'door', helper and intercessor for others. St Bernard speaks of her as an aqueduct between Heaven and Earth. The litanies depict the ways in which she helps those who trust her. The Virgin Mary has been the source of deep and extended popular devotion all around the world for many centuries (Warner, 1976).

Theologically both the Catholic and the Orthodox Church believe that Mary is the mother of God (*Theotokos*). According to the Council of Ephesus (AD431) Mary was 'the one who fabricates God in her bosom' as Jesus was incarnated in the Virgin through the action of the Holy Spirit. The Council also defined Jesus Christ's nature as completely divine *and* human: they are in a *hypostatic* union, a technical term used to describe the mystery of the nature of the second person of the Trinity (Jesus).

Space and movement in the sacred text

Space and the New Testament

In the New Testament, people are often on the move. Jesus and his followers are permanently on the road: they walk from village to village around the Sea of Galilee preaching the gospel. When the timing of religious festivals requires it, they go up to the centre of the Jewish world at the time, the Temple in Jerusalem. Sometimes they retire to solitary places such as mountains and deserts for prayer. They frequent specific buildings such private homes, mostly for meals, but also for ministering to the sick – and they sometimes teach in the small synagogues in villages. Buildings and spaces of power, such as the Jewish Temple in Jerusalem and Roman palaces and law courts, appear later in the New Testament story: as places of judgement, imprisonment and execution, first in Jerusalem and eventually in Rome.

Some spaces acquire special status by virtue of the events that happened in them, such as Mount Tabor (the site of the Transfiguration of Jesus), the Upper Room (the *Cenacle*) (the site of the Last Supper), the Mount of Olives and Golgotha. A woman, Egeria, is the first person to have left written testimony, in the late fourth century, of her long pilgrimage to the sacred places of Christianity, including the Chapel of the Burning Bush in Sinai (McClure and Feltoe, 1919). Across the centuries, these sacred sites became places of pilgrimage. Sacred places of pilgrimage continued to be established in the following centuries, often associated with apparitions of the Virgin Mary as at Lourdes and Guadalupe (Warner, 1976).

Comparison with the Old Testament: a people on the move

In the New Testament, as commented upon above, the spatial presence of Jesus and his disciples was characterised by being on the move from place to place. An important spatial dimension of Catholicism is the pilgrimage journey (Maddrell et al., 2015). Space as a path to and as a metaphor for something not physical in the New Testament is in many ways prefigured in the Old Testament that Christianity shares with Judaism.

Many centuries before Jesus was born, Abraham, Jacob and their descendants received the revelation of God being One, and not comprising the many gods of the various contemporary cults. Abraham was told by God to get out of his own country, and go to the Promised Land He would show him (Genesis Ch. 12). The people of Israel gradually gained a national identify, as a result of the pressures of abandoning their land of origin in Mesopotamia. They wandered for many years as a nomadic people, then going through slavery, Exodus, further deportations, exiles and wars, across the generations, before settling and building in the 'Promised Land'. Significantly, the first piece of land ever acquired as property by these nomadic people, many centuries before they were able to settle in, was to bury a woman – Sarah, wife of Abraham (Genesis Ch. 23, Ver. 1–20), in the field of Ephron, in Machpelah near Hebron in Ancient Israel (but now part of the West Bank territory).

Place as path and non-place

In both the New and the Old Testaments, cities are often not just physical sites but they provide analogies and allegories for spiritual 'locations' too, especially Heavenly locations. Looking at the range of spatial references in Scripture, one will find phrases that require theological interpretation, such as 'places to come', 'the house of my father', 'the Kingdom' (which is not of this world) and the New Jerusalem and Celestial City. The New Jerusalem, as described in the Book of Revelation (Apocalypse), is the 'place' of 'eternal life' where believers live in the love and peace of God. Metaphorically described in physical terms with measurements, orientations, materials and colours, in the Book of Revelation (Ch. 21, Ver. 12–27), this 'city' should be considered an early predecessor of the idea of the utopian cities (literally non-place cities). Spiritually speaking, the Kingdom of God is not necessarily contained in an actual physical city (*urbs*), nor intended to provide a model for social or political organisation (*civitas*) on this Earth, although the concept of an 'ideal city' has inspired many philosophers.

Saint Augustine in his book *The City of God*, of AD426, set out the design of the perfect city (Augustine, 2004), and Thomas More wrote *Utopia* in 1516 (More, 1989) on the ideal urban society. Later secular utopian thinkers all contributed to the development of modern city planning well into the twentieth century. The notion of a future ideal city still fuels the imagination of secular city planners as a recurrent idea within Western thought. But the Christian utopias were much more to do with metaphorically creating a spiritual celestial city, part of the Kingdom of God, where the evils of present-day society would disappear. We will return to the theme of utopias in the last section of this chapter.

The role of women in the Church

More women than men

The Church, strictly speaking, is not only an institution, but comprises the whole community of the believers, and is a manifestation of the Kingdom of God. Women comprise the majority of Catholic Church members, approximately 1.272 billion (Pew Research Center, 2015). Historically they have played, and continue to play a very significant role in many different dimensions of the life of the Church. According to L'Osservatore Romano (2016), the number of professed women religious (that is nuns and other consecrated women) worldwide in 2014, was 683,000, which is significantly higher than the total of religious men, including priests (415,792), permanent deacons (44,566) and consecrated lay men who are not priests (54,559).

The historical role of consecrated women in the Church

The New Testament provided a breakthrough for women by the standards of the time, believers being seen as neither male nor female but equal (Galatians Ch.3, Ver. 28) in God's eyes. But right at the beginning of the Old Testament it states that God created 'man' (mankind), 'male and female he created them' in His 'own image and likeness' (Genesis Ch. 1, Ver. 27). So woman is created in the likeness of God too. Such Jewish and Christian principles have, arguably, fed into Western culture over the centuries and contributed to our contemporary understanding of the principle of gender equality.

For many centuries, the Roman Catholic Church opened up opportunities for women that civil life could not provide, through the alternative that a consecrated life provided outside of marriage (Duby, 1990; Martínez, 2000). The

Church has historically provided important opportunities for leadership to women of any social class and background. Since its early times, the Church created possibilities for a life in community with other women, independent from the authority of husbands, fathers or other male relatives. Only women from the higher echelons of the aristocracy, and later the bourgeoisie, would have both the economic means and the legal status that allowed for an independent life out in the world. Convents were, from the fourth century and up to the twentieth, an alternative life choice for unmarried and widowed women, particularly for those without significant independent means of existence, whose fate was poverty.

Men and women: equal in dignity, complementary in vocation

Under Christianity women and men are fundamentally equals in dignity and value, but at the same time, the Catholic Church gives them specific and different vocations in the world. The understanding of gender equality in Catholicism falls within what has been called a 'feminism of difference' that recognises dissimilarities among men and women. Women and men are seen as having complementary roles (John Paul II, 1988). But at times, in the long history of the Church, the focus on complementarity and difference has been overemphasised, with significant negative social impacts. This reduced women's chances of accessing education and employment, and blocking them from entering into non-traditional feminine types of employment, such as the technical and scientific professions.

An understanding of contemporary Catholic perspectives on women's roles can be found in the work of philosopher Edith Stein (1996; 2015), a convert to Catholicism (from Judaism) who was canonised as Saint Teresa Benedicta of the Cross and is patron saint of Europe. Stein, a feminist in her youth, supported the radical defence of women's equality in education, the professions and the workplace, while setting this within a Catholic conceptual framework based on the 'feminism of difference'.

Writing in the early twentieth century, Stein's discussion on women's professional vocations supports the full professionalisation of women in any and all fields of endeavour, giving women a choice, which was a quite radical, even revolutionary position at the time. She defined a specific feminine professional vocational *ethos* (character) which could manifest itself through motherhood, chastity or traditionally feminine professions, but also if women are so led into (what at the time were) traditionally male professions.

Stein's discussion goes on to analyse how and in what ways this varied choice comes about, the specific implications for individual women, and the consequences for the nature of the workplace, the professions and society itself. Important sections of Stein's discussion are particularly relevant to contemporary issues, such as initiatives to attract more women into male-dominated STEM subjects (including science, technology and engineering), Stein's concept of ethos, and the possibility of women having a clear vocation to enter erstwhile male-dominated professions, is particularly relevant to the built environment professions (a field where efforts to attract women students and to keep women professionals have not always been successful). Overall, Stein's position offers a reasoned foundation to balance arguments for radical equality in the workplace, while still valuing traditionally feminine roles in the lives of Christian women entering the secular sphere.

Male priesthood and the sacrament of ordination

Women are increasingly encouraged to take leadership roles in the Church, but the priesthood is not open to them. Contemporary aspirations towards radical gender equality in public life and in the workplace, as well as political and social demands for female equality, are often seen to be at odds with the Catholic principle of a male-only priesthood. The established hierarchical and clerical structure of the Vatican prevents women from reaching positions of responsibility and leadership because the positions are only open to ordained men.

The argument for supporting male-only ordination is theological. Unlike in Protestantism, the office of the priesthood is considered to be a sacrament (compare with Chapters 4, 6, 11 and 13 whose authors all accept women clergy). Sacraments are God-given, and hence not subject to earthly authority or choice. We are entering deep theological waters. But the arguments may be summarised as follows (Catechism, 1993: Article 6; Ratzinger, 2017). Sacraments are essentially a symbolic representation, an outward sign of an inward and hidden reality. They are not practical tools designed by the Church, but are given by God. So the Church cannot change them but can only faithfully transmit them as the composition of the sacraments falls outside the limits of the Church's authority.

Male ordination is based on the revealed word of God, which shows that Jesus only appointed men to be his disciples (the 12 apostles). The manhood of Christ and of the apostles to whom he conferred the exercise of this ministry is neither secondary nor accidental, but an intrinsic component of the sacrament over which the institution of the Church has no authority. The proposition that Jesus only appointed men because of the customs of the time is not considered to be an acceptable 'get out clause' by the Catholic Church. Since Jesus already gave so much attention to women, in contradiction to the local customs and prohibitions of the time, he could have appointed women too, if he had wanted too. After all, priestesses were to be found in many cults at the time. But he chose not to do so.

Priesthood within Catholicism is not seen as a profession, or as a job. It is a vocation or state of life, to be exercised as a service that provides the means to help others in the path of sanctification for believers. Receiving the sacrament of orders (being ordained, becoming a priest) is not a right but a gift. Whoever feels a call for it must submit to the authority of the Church which has the responsibility and the right to call someone to receive it.

Those ordained in the priesthood of Christ are acting *in persona Christi Capitis* (in the person of Christ). This means that Christ himself is present in the person of the priest as 'the only mediator between God and men' (1 Timothy Ch. 2, Ver. 5; Catechism, 1993: para. 1544): as 'a high priest according to the order of Melchizedek' (Hebrews Ch. 5, Ver. 10 and Ch. 6, Ver. 20; Catechism, 1993: para. 1544). Ordination confers the priest with 'sacred power' so that when he ministers the sacraments it is the grace of the Holy Spirit that is involved, irrespective of the personal merits and sins of the human being who is the priest (Catechism, 1993: para. 1544).

The most important of all the sacraments in which Catholic priests act *in persona Christi Capitis* is the celebration of the Eucharist (the Mass) (Pope Paul, 1964). The liturgical rites celebrated today were first instituted in the second century. For Catholics, as well as for Orthodox believers, and unlike most Protestant churches, the Sacred Sacrament of the Eucharist is seen to contain the body of Christ in a real, true and substantial way, as well as his soul and divinity, and hence Christ himself in real and integral presence (Catechism, 1993: Article 3). This presence is not experienced through the senses but by faith. The consecration of bread and wine by the priest results in the conversion of the very substance of the bread (the Host) into the substance of the body of Christ, and the wine, into his blood. This mystery is called the transubstantiation. In the celebration of the Mass, Christ is made present and real in both the person of the priest and in the sacred form of the Eucharist.

In addition to the administration of the sacraments priests exercise leadership over their congregations. But this form of leadership should be seen as that of a shepherd looking after his flock. It should not be compared with secular power-centred and dominating forms of leadership but understood rather as one of service. It is based upon a model of 'servant leadership', exemplified by Jesus the Rabbi washing his disciples' feet at the last supper as narrated in John's gospel (John Ch. 13, Ver. 1–15).

Leaving the theological arguments aside, and speaking the secular language of policy, it could be argued that male-only ordination should not be considered a discriminatory practice. Rather it is a practice comparable to that adopted by secular governments when they appoint a particular type of qualified person or expert to work in a certain specialist field. In any case, the arguments that sustain the appropriateness of sex restriction are of a theological rather than sociological, economical, managerial or political nature.

Women in the institution of the Church today

While only men can be priests, today the vast operations of Church organisations in the fields of social work, education, missions and health, in parishes and dioceses, offer many opportunities for women to serve too. Women occupy all sorts of positions and roles. Some may still work in lower echelon jobs, but some occupy management, leadership and executive roles (Brennan, 2016). Most Catholic women, however, do not work for the Church itself but are 'in the world'. Catholic women work in all kinds of occupations and in all professions and at all levels, many at the highest ranks of their professional fields. For many, their Christian faith is a motivation in their careers.

For many women reconciling work outside the home with looking after their families represents a challenge that is not always easy to solve. Furthermore, within the Vatican, the many dioceses distributed around the globe, and in many ecclesiastical organisations, women are sometimes treated in patronising ways and as second-class religious. In these places, nuns fulfil all kinds of household tasks, from cooking, to cleaning, to washing, and are treated as domestic servants. All sorts of menial housekeeping tasks are still done by women religious who have completed theological studies. The first woman to work (in a non-domestic role) at the Roman Curia did so in the 1950s. By 2005, 21 per cent of the Vatican workers were women, a percentage that remained constant up to 2015 (*International Business Times*, 2015).

The way forward for women?

What does this all mean for contemporary demands for gender equality in the Church? First, in religious institutions theological issues come first, and political and administrative concerns, second. Thus the issue of increasing the participation of women in leadership positions still needs to be addressed with the proviso that it is accepted that the priesthood will continue to be male-only.

Some steps are being made to advance women in more senior (but non-priestly) roles. Since 2014 women have been appointed to several important positions of responsibility at the Vatican, including within the Pontifical Academy of Social Sciences and the International Theological Commission. Many positions that are open to male laity could in the future be occupied by women, such as diplomatic roles (for example at the United Nations), judges and magistrates, and economic managers. Guidance on reform from the so-called C9, a Council of cardinals set up by Pope Francis, states that only those positions that specifically require to be occupied by clergy will be so occupied, opening the door for the nomination of women. Other issues requiring attention, as pointed out by the Voices of Faith Group event in October 2016, relate to Church procedure and voting rights. For instance, women superiors of religious orders, who have the same canonical status as their male colleagues, are not allowed to vote at global Bishop Synods.

At a conference organised at the Vatican on Women's Day by the non-profit organisation Voices of Faith, Carolyn Woo, President of Catholic Relief Services, proposed the following recommendations (Voices of Faith, 2016):

- that leadership positions occupied by women become a normal thing instead of rare exceptions;
- that women are treated as full members in their own right, and not as invitees who are not to participate in decision making;

- that the Church takes an active role as advocate for the dignity and the rights of women against the many discriminations women face around the world, for instance on access to property, inheritance, or to public life;
- that these demands are listened to as a source of improvement for all, and not as a threat (i.e. to the male priesthood);
- that the idea of the feminine integrates in ample ways women's capacities and gifts, including perseverance, capacity to speak the truth even when it is not well received by those in power, and but also respect for more traditional 'female qualities' such as kindness, maternal instinct or aesthetic sensibilities (the sort of feminine qualities that contemporary secular feminist would define as gender stereotypes).

As the Church advances along these paths, new opportunities will arise for women to fully develop their gifts and talents, in all professional areas within the many institutions of the Church and at the higher echelons of the Vatican. The numbers of women in all sorts of policy sectors and the extension of the operations of the Church in many countries increases the chances of women shaping, or at least influencing, Church decision making. For example, women experts in architecture and urban planning would be ideal to give advice on church building and urban conservation issues. Many organisations within the Church hold very significant land and property assets, requiring this type of professional work as well as involvement in construction management, investment decisions and property law.

Revisiting utopia: women building communities

In this section, the aim is to take another look at 'utopia' with particular reference to women's ideas and communities. As will be seen, women have contributed historically in different ways to building Christian communities, sometimes, as leaders of groups of both men and women, at other times by living together among them, without men or families. They have created communities dedicated to the service of others, or to prayer and study in the confines of the cloister.

Women in the early communities

Early Christians created communities in which people were able to develop new kinds of personal and spiritual bonds, based on Jesus' parting instructions to his disciples to: 'love one another as I have loved you' (John Ch. 13, Ver. 34). The word 'church', in the original Greek *ekklesia*, literally means assembly, a group of persons and a community. The Church is the community of the believers. According to Acts Ch. 2, Ver. 44, at the beginning of the Church's creation, believers came together and 'held all things in common' and shared whatever material wealth they had with the poor, orphans and widows. Acts Ch. 6, Ver. 1–7 describes how persons of repute were appointed for the specific function of ensuring bread was equally distributed among the widows irrespective of them being of Jewish or Greek origin.

New community groups continued to be created as Christianity expanded in the first century, where believers could share communion with other Christians and in the communion of the Holy Spirit. Such groups were often more important to their members than their own families. But, as yet, the Christians had not set up separate religious communities with their own buildings. Lay communities would meet for collective prayer at private homes, often headed by women, at times behind closed doors to protect believers from persecution. Such tightly knit, home-based community groups of spiritual practice were extremely successful. They spread rapidly throughout the Roman Empire because of the novel protective social and economic networks they created preventing many members, particularly women, from being ostracised by society for their beliefs, and falling into destitution.

The order of the widows and the virgins

In the early Church, a variety of different 'orders' were gradually established, for example, for those ordained as priests, for married believers, for widows and for virgins. Each of these groups, or bodies, was an integral part of the Church, the body of Christ. While the Acts of the Apostles and the books of the Corinthians in the New Testament hint at a range of ministries, 'orders', gifts and vocations, greater information is to be found in the Apocryphal writings, which comprised various texts produced during a period spanning from the late first century up to the early third century. Some such books and epistles were subsequently included in the official Apocrypha, and are part of the Catholic Bible. Other writings, known as the *Pseudepigrapha*, did not make it into the official canon. Davies (1980) suggests that some of these books were written by women for other women. The audience would be either widows or women who after being converted to Christianity had abandoned their husbands for the Church, choosing a life of perpetual celibacy. A recurrent theme regarding the women featured in these texts is their problematic life experiences, and the difficulties they encountered when, for the sake of their new faith, they abandoned their husbands, some of which were powerful pagan civic and political leaders. According to this hypothesis, and as described in the texts, these women revolted against their husbands and their marriages, against marriage in general, and against the basic social patterns of the time, while, in the process, creating feminine celibate and ascetic communities.

The widows, a term that included both widowed women and those who had left husbands, who had adopted vocations of perpetual celibacy, became an important group or order. Their spiritual and charitable roles were acknowledged and subsequently defined by the Church authorities, and by the end of the fourth century 'the widows' movement had merged with the beginnings of feminine monasticism (Duby, 1990). Likewise, young women who decided not to marry but remain virgins and live the celibate life for the sake of the Kingdom remained all their lives inside the Church, with the economic and social support of the Church community, as part of the order of consecrated virgins.

The cloistered life of feminine monasticism: ora et labora

When in the fourth century Christianity became the official religion of the Empire, the Christian 'cult' left private homes, and churches were built (or pagan temples were remodelled), giving the Church an architectural presence in urban space. (The religion moved from the private to the public realm, from the home to the public square.) Women lost ground as the small communities gave way to the new hierarchical structure of an institutionalised Church, under the authority of an archbishop assisted by bishops and deacons, with clear distinctions between the clergy and the lay. However, women believers from the aristocracy still had significant informal power and influence in the workings of the Church.

As the Church grew in power, many yearned for the simple religious life of earlier times. Thousands of men and women left the cities in the fourth and fifth centuries to become hermits in the deserts searching for a life in aspiration for the 'celestial city'. This solitary way of life was soon reorganised by grouping people in cenobitic (simple monastic) communities of women and of men, opening the way for the development of the monastic organisations that became the backbone of Christianity.

Monasteries (and nunneries) provided the spatial and social organisation that allowed individuals to live as a community in ways conducive to the spiritual experience. Saint Benedict of Nursia wrote in the sixth century the rule by which cloistered monastic contemplative life has been organised in Catholicism ever since (Benedict, AD 516). Monastic architectural typologies were devised to provide the physical environment in which such life could be lived, often in remote places away from cities. Through the motto *ora et labora* (pray and work) Benedict organised the rhythm of life around the Liturgy of the Hours – the cloistered monks and nuns praying and singing seven times a day, between periods for work, leisure, common meals and personal prayer.

The monastic orders of women, and men, gradually became self-sufficient communities independent not only from secular authority but, to some extent, from the Church hierarchy. They owned very significant agricultural estates and economic enterprises, as well as building some of the best examples of Gothic architecture. Women's monasteries (including nunneries and convents) were initially created as off-shoots of or dependent on men's organisations. The creation of the first totally independent female monasteries was not without difficulty, as in the experience of Saint Hildegard Von Bingen in the twelfth century (Maddocks, 2013). Schulenburg (2005) describes attempts by the male Church hierarchy to control women's monastic organisations, restricting their ministry among the general public and within the city, by containing them within enclosed orders. But with time it was accepted that local bishops and superiors had no power over these communities, as many were put directly under protective papal authority, giving them the choice of being open or enclosed orders.

At the female monastery, women could follow a life dedicated to prayer and contemplation, and undertake a variety of artistic, managerial and political and scientific endeavours in the company of other women. The earliest orders, including the Benedictine, Cluny, Cistercian and Poor Clare Nuns, were mainly contemplative. Life at the convent was often very modest and sometimes on the verge of poverty. Some orders were mendicant (supported by begging). But, others allowed richer women of means to bring their wealth and possessions with them, financially benefitting the community but also creating internal social distinctions.

Within female monasteries, convents and abbeys, from the Middle Ages onwards, many women achieved positions of significant economic, social and political influence in society, managing extensive estates and organisations with hundreds of members extended sometimes across many countries, as abbesses, mother superiors and generals of religious orders, and as mystic counsellors to kings and popes. The most famous include Saint Catherine of Siena, Brigitte of Sweden and Hildegard von Bingen (Maddocks, 2013). But there are thousands of other female leaders, of various social classes, who over the years ran hundreds of religious congregations of sisters with specific vocations for service, including running hospitals, orphanages, schools, asylums and ministries to single mothers, prostitutes and the homeless, and working as missionaries throughout the world (Brennan, 2016).

Right up to modern times, women continued to be active in establishing new religious communities and creating charities and vocational specialisms to meet specific needs. A recent study has identified 100 such new foundations that were set up by women in Spain across the nineteenth century (Torres López, 2019), all of them still in existence and working. The women founders were great leaders who were able to raise the funds, obtain or build the buildings, get political support from public authorities and obtain the corresponding authorisations from the Church.

Urban communities of working semi-religious women: the Béguinages

The *Béguinages* (*Beaterios* in Spanish), were institutions comprising sets of buildings and sometimes entire neighbourhoods, many still existing today, in which unmarried and widowed women lived together – without men. Women in the Middle Ages could enter this form of semi-religious life which flourished particularly in the textile centres in Flanders and Germany, such as the one in Bruges (now a UNESCO heritage site) (Majérus, 1997). *Béguines* had non-permanent celibacy vows so they could leave to get married. And they could attain financial independence by working in the textile industry. Thus lay women could voluntarily escape the authority of husbands and fathers, the very real dangers to life of continuous pregnancies, and the hard labours of childrearing and housekeeping in premodern times to devote themselves to prayer, learning and charity (Swan, 2016).

The *Béguinages* were not enclosed, secluded communities, so women had liberty to move around the city, gaining respect and protection in wearing the habit (compare Chapter 8 on the protection of religious clothing). As semi-religious women, they may be considered to be the precursors of later 'lay' congregations of sisters. But, their presence and liberty of movement in urban space was not without opposition from the clergy, the guilds and the municipality. They were legally subject to certain municipal laws and duties. For example, the city could require them, at personal risk, to attend the sick in times of plague, as unpaid nurses. But professionally and architecturally, the *Béguinages* had an impact on cities; their building complexes designed by women constitute early examples of women and planning principles. The *Béguines* themselves also provide living examples of alternative social life paths and employment opportunities for women – outside of both the convent and marriage.

A feminist utopia: the City of Ladies

The concept of the ideal city in which the evils of present-day society would disappear is a recurrent theme. This is illustrated in Christine de Pizan's fifteenth-century publication, *The Book of the City of Ladies* (Pizan, 1405). This may be the first feminist urban planning book. Pizan reveals and confronts the many prejudices and injustices perpetrated against women in her time. She tells the stories and achievements of many famous women throughout history. These women are housed in her city of which they are also the building blocks. In a fictitious scenario, her heroines, drawing on their virtues and merits, debate with famous contemporary male philosophers, and successfully argue in favour of women's education and participation in public life.

Pizan's book included ideas on how to design cities, to create spaces for both women and men that challenged ancient gendered zoning principles, which generally supported women being restricted to domestic roles. For example, in her 'dream vision' of the ideal city, she envisaged high buildings with towers and turrets, surrounding welcoming courtyards inside the walls where women could read, discuss philosophical and religious ideas and pursue their education.

Lay initiatives: intentional communities

Lay women, as well as religious sisters, were active in building new utopian ventures. Communities of minority lay populations built on common spiritual practice and shared principles for ordering everyday life have been created since the eighteenth century, especially within some parts of the Protestant world, particularly in the USA. Hayden (1977) and Kanter (1972) provide detailed accounts of these historical communities, some of which are still in existence, as to how religious and communitarian principles were manifest in urban planning and architectural design principles.

These communities show various degrees of integration with the surrounding city and society. They offer non-traditional alternatives for women and the organisation of family life, supported by communal co-operative housekeeping and shared childcare. As the medieval examples of the Catholic *Béguinages*, the convents and the monasteries, these more modern model communities are earthly utopias built for a life which embodied 'Kingdom' principles.

But ordinary lay women have always also had an important, but more mundane role at the urban neighbourhood and parish level in supporting the Church and laity. For example in Spain, urban space and daily living were historically organised around the local church. Parishes were places in which laywomen deployed their organisational abilities and contributed their voluntary work in support of pastoral, liturgical or philanthropic activities (compare Chapter 10 on women's 'invisible role' in municipal spirituality). Women religious have also always been involved at the local level, running schools, working in hospitals and setting up a range of missions and initiatives to deal with the problems of urban society and stem the tide of secularisation.

Rather than battling on trying to minister to and change modern society from within, some Catholics are taking the view it is better to start again. Currently the Benedict option (Dreher, 2017) is being developed as a social model for building alternative future urban settlements by lay Catholic visionaries. These will nurture the spiritual life of the laity in a post-Christian society where believers are becoming a minority. The Benedict option provides an adaptation of the monastic rules created by Saint Benedict, for lay people living in cities. The objective is to create a strong supporting framework to facilitate spiritual practice, ordering the different aspects of life towards sanctification. This would be for both men and women, and families and would continue to promote the 'Kingdom' on Earth.

References

Augustine (2004 [AD 426]) *City of God*, London: Penguin Classics.

Benedict, St (2008 [AD 516]) *The Rule of Benedict*, London: Penguin, translated and edited by C. White.

Brennan, R. (2016) *The Global History of Catholicism*, London: Rowan, Lord Brennon Catholic Educational Trust.

Catechism (1993) *Catechism of the Catholic Church*, Vatican City, Rome: Libreria Editrice Vaticana. www.vatican.va/archive/ENG0015/_INDEX.HTM (accessed 11 June 2019).

Davies, S.L. (1980) *The Revolt of the Widows: The Social World of the Apocryphal Acts*, Carbondale, IL: Southern Illinois University Press.

Dreher, R. (2017) *The Benedict Option: A Strategy for Christians in a Post-Christian Nation*, New York: Sentinel.

Duby, G. (1990) *Historia de las mujeres. Vol. 1 La antigüedad, Vol. 2 La Edad Media, Vol. 3 Del renacimiento a la Edad Moderna, Vol. 4 El siglo XIX* [History of Women. Vol. 1

Antiquity, Vol. 2 The Middle Ages, Vol. 3 From the Renaissance to the Modern Age. Vol. 4 The Nineteenth Century], Madrid: Santillana.

Fee, G. and Stuart, D. (1981) *How to Read the Bible for All Its Worth*, Grand Rapids, MI: Zondervan.

Hayden, D. (1976) *Seven American Utopias: The Architecture of Communitarian Socialism*, Cambridge MA: MIT Press.

International Business Times (2015) *Women's Day 2015: More Women Employed at the Vatican, but Only Two Hold Top Positions.* www.ibtimes.com/womens-day-2015-more-women-employed-vatican-only-two-hold-top-positions-1839370 (accessed 11 June 2019).

John Paul II, Pope (1988) *Apostolic Letter Mulieris Dignitatem* [The Dignity of Women]. https://w2.vatican.va/content/john-paul-ii/es/apost_letters/1988/documents/hf_jp-ii_apl_19880815_mulieris-dignitatem.html (accessed 11 June 2019).

Kanter, R.M. (1972) *Commitment and Community: Communes and Utopias in Sociological Perspective*, Cambridge MA: Harvard University Press.

L'Osservatore Romano (2016, April). www.osservatoreromano.va/en/news/living-church-changing-world (accessed 11 June 2019).

Maddocks, F. (2013) *Hildegard of Bingen: The Woman of Her Age*, London: Faber.

Maddrell, Avril, della Dora, Veronica, Scafi, Alessandro and Walton, Heather (2015) *Christian Pilgrimage, Landscape, and Heritage: Journeying to the Sacred*, New York: Routledge.

Majérus, Pascal (1997) *Ces femmes qu'on dit béguines: guide des béguinages de Belgique: bibliographie et sources d'archives* (Introduction bibliographique à l'histoire des couvents belges antérieure à 1796), Brussels: Archives générales du Royaume.

Martínez, C. (ed.) (2000) *Mujeres en la historia de España. Enciclopedia biográfica* [Women in the History of Spain: Biographical Encyclopaedia], Madrid: Planeta.

McClure, M.L. and Feltoe, C.L. (eds) (1919) *The Pilgrimage of Egeria*, London: SPCK (Society for Promoting Christian Knowledge).

McNamara, D. (2009) *Catholic Church Architecture and the Spirit of the Liturgy*, Mundelein, IL: Liturgical Institute, Hillenbrand Books.

More, Thomas (1989 [1516]) *Utopia*, Cambridge: Cambridge University Press, translated and edited by George M. Logan and Robert M. Adams.

Mumford, L. (1965) *The City in History*, Harmondsworth: Penguin.

Paul VI, Pope (1964) *Apostolic Letter Lumen Gentium*. Vatican City: Libreria Editrice Vaticana. www.vatican.va/archive/hist_councils/ii_vatican_council/documents/vat-ii_const_19641121_lumen-gentium_en.html (accessed 11 June 2019).

Pew Research Center (2015) *Demographic Profiles of Religious Groups*. www.pewforum.org/2015/05/12/chapter-3-demographic-profiles-of-religious-groups/ (accessed 11 June 2019).

Pizan, Christine de (1983 [1405]) *The Book of the City of Ladies*, New York: Persea Books, translated by Earl Jeffrey Richards. Also available at: www.academia.edu/10130771/The_Book_of_the_City_of_Ladies (accessed 11 June 2019).

Ratzinger, J. (2017) *Obras completas* [Complete Works], Madrid: Biblioteca de Autores Cristianos.

Sánchez de Madariaga, I. and Roberts, M. (eds) (2013) *Fair Shared Cities*, London: Ashgate.

Schulenburg, J.T. (2005) 'Gender, celibacy and proscriptions of sacred space: symbol and practice' in V.C. Raguin and S. Stanbury (eds) *Women's Space: Patronage, Place, and*

Gender in the Medieval Church, New York: State University of New York Press, 180–205.

Stein, E. (1996) *Essays on Woman*, Washington, DC: Institute of Carmelite Studies Publications.

Stein, E. (2015) *Obras completas. The Complete Works of Edith Stein*, 5 vols, Madrid: Editorial Monte Carmelo.

Swan, L. (2016) *The Wisdom of the Beguines: The Forgotten Story of a Medieval Women's Movement*, New York: BlueBridge Publishing.

Theresa of Avila (Teresa de Jesús), St (2008 [1577]) *Interior Castle*, New York: Dover Thrift Editions.

Torres López, J. (2019) *Movimiento fundacional de instituciones religiosas femeninas españolas en el siglo XIX. Pervivencias y cambios* [Foundational Movement of Feminine Religious Institutions in Spain: Permanence and Changes]. PhD dissertation, Madrid: Universidad Complutense de Madrid. https://eprints.ucm.es/51687/1/T40955.pdf (accessed 11 June 2019).

Voices of Faith (2016) *Intervention of Carolyn Woo at 8th March Event at the Vatican*. www.youtube.com/watch?v=vulMZqD5NrM (accessed 11 June 2019).

Warner, Marina (1994 [1976]) *Alone of All Her Sex: The Myth and Cult of the Virgin Mary*, London: Picador.

4 Spaces of interfaith dialogue between Protestant and Muslim communities in Germany

Petra Kuppinger

Abstract

This chapter examines moments and spaces of a year-long interfaith and cultural dialogue series in an urban neighbourhood in Germany. Based on ethnographic research in Stuttgart, I illustrate that the articulation of urban spiritual geographies is not limited to prominent nodes of the urban religious geography, but crucially depends on many small spaces, seemingly invisible exchanges, and interpersonal encounters that create more inclusive spiritual geographies that are reflective of larger urban religious and cultural transformations. I argue that it takes more than the dissemination of information and knowledge for interfaith and intercultural encounters to be successful. Meaningful interfaith encounters need to undo invisible boundaries and create feelings of familiarity and community among diverse local residents. Productive interfaith activities need to become starting points for new ties, networks and future cooperation, remake urban spatialities and socialities, and generate more inclusive cityscapes.

Key words: Islam, Christianity, Stuttgart, Germany, interfaith dialogue, spaces

Faith groups coming together

In this chapter I examine moments and spaces of this interfaith and cultural dialogue series. Based on ethnographic research in Stuttgart, I illustrate that the articulation of urban spiritual geographies is not limited to prominent nodes of the urban religious geography, but crucially depends on many small spaces, seemingly invisible exchanges and interpersonal encounters that create more inclusive spiritual geographies that are reflective of ongoing urban religious and cultural transformations. I show that it takes more than the dissemination of information and knowledge for interfaith and intercultural encounters to be successful. Effective interfaith encounters need to undo invisible boundaries, create feelings of familiarity and community among local residents, become starting points for new ties, networks and future cooperation, remake urban spatialities and socialities, and generate a more inclusive cityscape.

On a Thursday evening in March 2011 about 20 women and men were seated around a conference table in a Protestant community centre in Stuttgart, Germany. All eyes were on Mahmood Alwani who lectured to an audience of Muslims and Christians about humour and its role in Islam (all names unless otherwise indicated are pseudonyms). Some seemed familiar with aspects of this topic. Others were new to the subject. Regardless, everybody listened to Mr Alwani, a young scholar who had recently completed his Master's degree in Islamic studies at a local university. Mr Alwani explained the role of humour and laughter in the Qur'an and Sunna. He discussed some *hadiths* (accounts about the life, acts and sayings of the Prophet Muhammad) and how the Prophet explained the role and use of humour among Muslims. After these theological, historical and theoretical elaborations, Mr Alwani described the use of humour in popular religious traditions and everyday life. He introduced Nasreddin Hoça (in Turkish) or Goha/Joha (in Arabic), who is the central character of many humorous and also pedagogical stories/jokes. It seemed that the audience had been waiting for the 'applied' part of the presentation, because quickly questions about more 'real' life examples emerged. Could Mr Alwani tell more humorous stories or jokes, for the non-Muslims to get a better glimpse of Muslim humour, and for everybody to have a good laugh? Mr Alwani obliged and told more Hoça/Goha tales/jokes and the audience had their laughs. The event became increasingly relaxed and informal. Participants added similar funny little stories. Egyptian political humour, comedy and Saidi (Upper Egyptians, the butt of much joking) jokes were mentioned and examples told. In the course of almost two hours, the event turned from a lecture to a casual conversation among people with a shared interest. Laughter punctuated the exchange.

Mr Alwani's lecture was part of a year-long interfaith dialogue and cultural series entitled 'To Approach One Another' (*Aufeinander Zugehen*) that had been organised by the local Protestant church and mosque communities in Zuffenhausen, one of Stuttgart's most diverse neighbourhoods. The series included 33 events in local venues that featured diverse speakers and presenters. Mr Alwani's presentation was one of the last events in the series. Some participants had become 'regulars' of the series, others were casual visitors. The shared laughter had pulled in everybody and created a sense of closeness that rarely occurs at more formal events. The shared laughter created a momentary bond among participants. The event was more relaxed than others in this series, and the room turned into a comfortable home space. Some visitors noted at the end of the evening that this was one of the best events of the entire series.

Starting in March 2010, the Protestant Church and the mosque communities in Stuttgart-Zuffenhausen ran an ambitious 12-month interfaith and intercultural dialogue marathon which included a remarkable variety of events (lectures, music, performances, food) and presenters (clergy, politicians, scholars, journalists, lay people/members of religious communities), and engaged diverse topics (faith, everyday practices, politics, popular culture, migration) and used many local venues. The programme was designed to enhance communication and social relations between Muslims, Christians and others in the quarter, but also attracted visitors from elsewhere in the city. In addition to interfaith dialogue, the series aimed at providing spaces of cultural exchange and learning. It included local presenters and others from Stuttgart and the region. Beyond providing opportunities for individuals and communities to learn more about the other religion, the series brought people together, created new ties and relationships, produced shared experiences, seeded future cooperation and remapped individual and local spiritual and cultural geographies.

In minute encounters that include a multitude of actors and participants, activities and many local spaces, urban cultural, religious and emotional geographies are carefully negotiated. To understand transformations of urban religious, religiously inspired and emotional geographies, it is crucial to look at small spaces where religious lives and religiously inspired interactions unfold. Analysing features of the 'To Approach One Another' series, I illustrate how elements of faith, communication and neighbourliness turned this programme into a vibrant platform of religiously inspired civic engagement, interaction and change. The combination of topics, venues, presenters and visitors neatly framed not just isolated moments of interfaith dialogue, but outlined discursive and material spaces and possibilities where different constituencies could feel comfortable at present and in the future. When participants met in local venues, spaces were (re-)mapped as welcoming or home spaces to constituencies that had not previously visited these spaces. Using different localities enhanced the map of dialogue and engagement between local groups. In the course of the year, the programme levelled boundaries, created new links, remade individual and communal mental maps and the neighbourhood's religious and cultural geography, and seeded foundations for continuing cooperation. The remapping of local spatial imaginaries and material spaces, the inclusion of Muslims into a broader array of spaces and the insertion of Muslim spaces into the neighbourhood's religious and cultural geography are vital aspects of resulting transformations of urban religious landscapes.

Cities and lived religion

Recent decades witnessed the emergence and expansion of urban religious and cultural studies, the study of urban spiritual geographies and the analysis of post-secular cities (Orsi, 1999; Tweed, 1997; Kong, 2001; 2010; Beaumont and Baker, 2011a; 2011b; Cloke and Beaumont, 2013). Lily Kong insisted that religious geographies are part of larger urban and spatial dynamics and added that 'the politics of religious spaces are also tied up with gender, race, and class politics, and politics between nations' (2001: 217). Scholars maintain that religion and religious transformation are not isolated urban features, but are reflective and constitutive of urban cultures and transformations (Hervieu-Léger, 2002; Desplat and Schulz, 2014). In the context of globalisation and global migration, migrant and/or minority religions have considerably enlarged, diversified and transformed religious geographies in global/globalising cities (Stepick et al., 2009; Livezey, 2000; Dwyer et al., 2016; Becker et al., 2014). Studies emphasise the relevant roles of religious actors in global cities and how their 'embodied and spatialized practices and modes of belonging and exclusion' matter in the complex 'local presences and global flows' in their urban environments (Garbin and Strhan, 2017b: 10, 2). In the context of Europe, many scholars have analysed the localisation and civic participation of Muslims and their communities (Werbner, 2002; Bowen, 2007; 2010; Kuppinger, 2015a). Peggy Levitt (2008) highlights the often neglected function of religion in public life and civic interactions and argues that religion, faith-based activities and faith-inspired actors need to be taken more seriously by analysts. She explains that for many immigrant/minority religious groups, membership in a faith community does not translate into isolation, but can 'socialize[s] members into receiving country politics' (2008: 768), while simultaneously allowing believers to maintain ties to home countries. Jürgen Habermas' (2006) reflections on the role of religion in the public sphere and how it can be understood and accommodated highlight the critical role of religion in the (urban) public sphere. These interventions outline the vital civic role of religion and argue that immigrant/minority faith communities often create venues or modes of civic engagement for their members.

Engagement in minority religious communities can teach individuals valuable lessons about civic participation and expose them to new constituencies, modes of exchange and patterns of interaction. Interfaith activities introduce individuals to new urban platforms and create links and networks. Minority religions' places of meeting and worship in global cities play crucial roles for members and their participation in the larger urban religious geography and civil society (Garbin and Strhan, 2017a). These houses of worship or community centres are home spaces for new arrivals and comfortable points of momentary withdrawal for long-term urbanites (Fesenmyer, forthcoming). They are also stepping stones for participation in urban civic and cultural spheres. In sheltered and more homogenous houses of worship or faith-based groups, individuals – and here especially also women – often learn important organisational, social and cultural skills which they then can then apply in other civic contexts (Kuppinger, 2015b).

Immigrant/minority houses of worship or community centres frequently remain materially and symbolically invisible as popular resentment combined with limited financial resources push them into remote, often commercial or industrial, quarters where they rent or buy defunct industrial sites or empty warehouses (Garbin, 2013; Kuppinger, 2011). Having a home space constitutes a critical starting point that allows communities to grow roots and articulate local practices, religiosities and religious lifestyles, develop a sense of home in the city and cooperate with others (Azzara, 2019). It allows communities to invite others and participate in events that require reciprocity (Kuppinger, 2017). In the process of everyday activities and involvements, religious subjectivities, practices and subcultures emerge that are locally grounded and globally linked. Minute moments of cooperation, participation and interaction weave communal spaces and community members into the larger urban religious and cultural fabric. Yet, even when communities grow robust roots in their immediate environs and interact with neighbours and civic associations, their efforts often remain invisible for the urban public which remains reluctant to recognise these communities. It takes much effort for minority religions, especially Muslim communities, to become visible and respected by dominant society.

For interfaith and intercultural encounters to contribute to the making of egalitarian urban spiritual geographies and inclusive cityscapes, it is not enough to disseminate information among different constituencies. Encounters need to undo invisible boundaries and create a sense of familiarity, welcome and shared interests and experiences. Participants in productive interfaith dialogue events need to see, hear, feel, experience and indeed taste elements of unfamiliar religious and cultural contexts. These mundane and affective elements might outweigh informational ones. For spaces and moments of dialogue to become agents of urban cultural change and mediators for more inclusive cityscapes, they need to include social and affective seeds that become starting points for new ties, relationships, networks and future cooperation. More inclusive cityscapes are not constructed in municipal offices or board rooms, but are articulated in small spaces where ordinary people get to know each other and in tiny steps remake their sense of self and others and become familiar with new ideas and spaces. Sharing food, listening to music, discussing shared interests and becoming comfortable in new spaces, local residents remake urban spiritual and cultural geographies.

Stuttgart, Islam and Muslims

Stuttgart is the state capital of the southwest state of Baden-Württemberg and the sixth largest city in Germany (600,000 inhabitants). Metropolitan Stuttgart (including five additional counties) has more than 2.5 million residents. Stuttgart is a high-tech, car and banking city. In the competition of German cities, Stuttgart scores with its global industries (e.g. Mercedes, Porsche and Bosch) and growing banking sector (second after Frankfurt). Unemployment rates are among the lowest in Germany, and social programmes and projects receive substantial funding. Stuttgarters experience considerable differences in wealth, income and size and quality of housing, but these differences are less pronounced than in other Germany cities, and seem benign in comparison with other global cities. Stuttgart is a contender in the race of (secondary) global cities, as the city seeks to attract investment, tourists and globalised events and spectacles. In recent decades the new Mercedes Benz Museum and Porsche Museum were constructed, and a sizeable fair and exhibition centre was added to the city's list of attractions.

Stuttgart has the largest share among German cities of residents who are either immigrants themselves or descend from post-war migrants to the city (what Germans call migration background, or *Migrationshintergrund*). In 2012, 39.9 per cent of Stuttgart residents had such a *Migrationshintergrund*. For those younger than three years, the figure was 57.5 per cent. About 20 per cent of the urban residents are foreign nationals (Landeshauptstadt Stuttgart, 2013: 12). In the daily lives of Muslims and their communities in the city, economic aspects play an important role, but everyday lives are not centrally marked by struggles for survival. Tensions are often cultural and political and concern questions of recognition, participation, civic rights and access to space. For example, conflicts about the construction of mosques are less about whether a congregation owns the funds to buy suitable real estate, but more about whether this real estate is made available for mosque construction (Kuppinger, 2014). Despite its considerable Muslim population – almost 10 per cent of the population, about 60,000 people (Baden-Württemberg, 2005: 10) – Stuttgart does not have a purpose-built mosque. There are about 25 mosque associations in Stuttgart, which are organised as registered legal associations (*Verein*). This organisational form is favoured by German law and gives the respective associations privileges, such as access to facilities and funding possibilities.

To approach one another

In an urban quarter with approximately 35,000 residents, the 'To Approach One Another' series was hugely ambitious. The programme (March 2010 to March 2011) consisted of 33 events, some of which were hosted by public institutions (the public library; secular community centres), and a commercial venue (a book store). The central institutions of the series, the mosque and Protestant church are within ten-minute walking distance from each other (and incidentally most of the other venues used in the series). The series was conceptualised as Muslim–Christian dialogue, but included references to the Jewish tradition (Jewish music performance; excursion to Mannheim including visiting a synagogue and a mosque). Topics were not limited to religion, but encompassed culture, politics and history. In October, Muhtaram Aras (not a pseudonym), by then a member of the city council and head of the Green Party faction who at the time was running for a seat in the state parliament (since 2016 she has been the president of the state parliament), lectured about the meaning of integration. A few days later the author Nilgün Tasman (not a pseudonym) read from her book *I Dream in German*. Neither event directly addressed religion, but engaged multi-ethnic urban lifeworlds. In May 2010, a local Italian ice cream maker and a Protestant minister talked to an informal audience at a communal breakfast in a Protestant community venue about the surprising intercultural history of ice cream and sorbet. The same Protestant minister on another occasion in the public library spoke about the presence of Arab words in the German language. The series included a theatre production about the experiences of a Turkish family between Germany and Turkey.

The series used ten neighbourhood venues. The choice of many premises (instead of only Protestant and Muslim community facilities) illustrates the organisers' understanding that to change religious and cultural geographies and individual mental maps, it is crucial to move beyond the confines of houses of worship and community centres. The use of multiple spaces demonstrates the awareness that religious and cultural spheres are intertwined and interfaith dialogue and faith-based encounters are not limited to houses of worship. In the process spaces were opened to diverse constituencies and inserted in individual and communal spatial imaginaries. The organisers recognised the necessity of addressing different constituencies. Events were designed for specific groups like school children, youth, women or the general population and sought to address as many different constituencies as possible. Organisers aimed to provide different activities and experiences which would permanently remake the local religious and cultural geography and create robust grounds for future cooperation.

The series departed from dominant models where people talk about each other. Instead people talked *to* each other. One refreshing feature of the series was that there was no 'Women in Islam' event/discussion, a topic that often too prominently features in much engagement (by outsiders) about Islam. Women were simply there as presenters and participants and nobody needed to explain that. Local Muslim, Christian and other women and men casually met, debated relevant topics and simply talked to each other.

The 'To Approach One Another' series was not an isolated effort of the organising parties, but part of ongoing cooperation. For years, the mosque has worked with the neighbourhood council, churches and civic associations (Kuppinger, 2015a: 156–192). Mr Serdar, the long-time mosque president, is the pivotal figure in this cooperation. It has long been his goal to position the mosque as one among many local houses of worship and a partner for the community council, other religious communities and civic associations. The mosque often participates in public events, like an 'open culture night event', where residents could visit local institutions, houses of worship, public places and commercial venues until late into a summer night. Every spring the mosque hosts a *kermes* event (communal festival and fundraiser) where tents are set up in the mosque's front yard and members, friends and neighbours casually sit together, talk and eat. The mosque participates in the Day of the Open Mosque on the German National Holiday (3 October) when mosques open their doors to the public. A few years ago, the mosque community marched in a local parade which was an unusual step for the mosque and local secular communities. The mosque also runs a food stall at the annual neighbourhood fair.

Events

In October 2010 Hakan Yilmaz, an engineer and representative of the mosque's regional association, gave an informal presentation entitled 'How We Live Our Faith – Tradition, Holidays, and Customs in Islam' in the prayer room of the mosque where tables had been set up to form a larger square.

Cookies and juice were served. About 25 people attended this event, some were Muslims, but the majority where Christians/non-Muslims. Mr Yilmaz started with an introduction to central Muslim holidays. Mr Yilmaz was less concerned to impart sophisticated theological knowledge, but tried to convey the details of everyday religious lives and practices. What does it mean to be a pious Muslim in Germany? What are some of the possible difficulties for pious German Muslims? What does it mean to raise Muslim children in Germany? He described how he and his wife explained to their son that even though he might sing Christmas songs in his kindergarten (German pre-school), he was not a Christian. Mr Yilmaz emphasised that he had no problem with his son singing the songs, but he tried to teach him that he was a Muslim and had his own holidays. These holidays, very importantly for children, came with their own gifts so the little boy was not at a disadvantage versus his peers, only the timing was different. Mr Yilmaz explained that being a Muslim never kept him away from non-Muslims or prevented him from having fun. He related that when he was younger he went out with his friends, some of whom would drink alcohol. He had no problem with this and neither had his friends. They appreciated his presence, because they could always rely on him to remain sober and drive home.

Mr Yilmaz's lecture quickly turned into a casual question-and-answer format where non-Muslims asked about Mr Yilmaz' daily religious life and practices. What did praying mean to him? Were all prayers the same? Mr Yilmaz noted that praying was a moment of connecting with God and of peace, away from the daily work and running around. He stated that prayers at different times and occasions 'counted' differently and that, for instance, prayers during the 'Night of Power', when the first part of the Qur'an had been revealed, counted more than daily prayers. This explanation caused surprise among non-Muslims. One man wondered whether this was like an account system where one piled up points. Did he pray at work? Mr Yilmaz said he tried to do his regular prayers at work and it was no problem. During short breaks when colleagues went outside to smoke he prayed and nobody minded. Where did he pray? He said that was no issue either. Close to his office was a small conference room that was rarely used. So he could almost always pray there. The audience continued this lively question-and-answer session with Mr Yilmaz who made the audience feel very comfortable to ask more questions. The atmosphere moved from a lecture to an exchange among acquaintances.

In December 2010, two young women, Ms Şayma Öztürk and Ms Zeynep Harun, members of the mosque community, presented and moderated an event announced as 'Lecture and Conversation: Prophet Muhammad', which they renamed to 'Muhammad: The Prophet of Mercy'. The event used the same set up as Mr Yilmaz' and participants sat at tables (with snacks) in the mosque's prayer room. The two young women (one with a university degree in education science and the other in sociology) introduced aspects of the life and teaching of the Prophet. They started with a historical introduction to the Prophet's era and noted that the moment of his birth indirectly symbolised the end of the Persian Empire and how Mecca, which had been an 'out of the way small town', was catapulted centre-stage to become the heart of the growing Muslim empire/world. They emphasised the Prophet's momentous role in Islam and how he serves as a role model (*Vorbildfunktion*) to individual believers by way of his exemplary works, acts and teachings as chronicled in the Qur'an and Sunna. Going back and forth between Ms Öztürk and Ms Harun, they provided more details of the Prophet's life, times, teaching and relevance in individual pious lives. They explained that today young Muslims were caught in an 'intellectual imbalance', as they were caught between conventional religious teaching (*ältere Bildung*) and their youthful enthusiasm for their religion and very personalised relationship with the Prophet. Once the speakers had set out the larger parameters, they moved to answer questions from the audience. Christian participants eagerly took up this opportunity and many questions centred on a comparison between the role and position of the Prophet Muhammad and that of Jesus in the Christian tradition. Much like at the event with Mr Yilmaz, a lively conversation ensued where participants felt free to ask questions which were answered in great detail by Ms Öztürk and Ms Harun and the occasional additional contributions from Muslims in the audience.

On a Saturday morning in late January 2011, Christian and Muslims women met in a Protestant community facility for a breakfast where they shared food and talked about 'questions of faith and life'. Without a lecture or planned programme, women came together, shared home-made food and casually met and talked. The only purpose was to sit together.

The final festive event of the series was a lecture by a well-known local journalist, Reinhard Baumgarten (not a pseudonym), who at the time was preparing to take over the Istanbul office of a local public radio station (from where he was still reporting in 2018). The event had been scheduled for late March 2011 in a Catholic community centre. In light of the ongoing regional upheaval and revolutionary movement in Egypt, the topic of Mr Baumgarten's talk was announced as 'Egypt: Lived Encounters between Muslims and Christians'. His presentation elaborated on recent Egyptian politics as a background to the ongoing revolt. Unlike at presentations described above, at this event the audience was seated in rows. A small buffet was set up in the hall. Mr Baumgarten introduced relevant actors and dynamics in Egypt and the region that helped to shed light on the ongoing struggle and upheaval. The audience listened to his historical and political elaborations as Mr Baumgarten explained elements of the revolt and the agendas of relevant politics groups. Unlike other events in the series, this evening remained formal and did not turn into a casual exchange (the speaker left soon after the end of his talk). Regardless, it pulled the audience together for different reasons. Muslims and Christians alike were interested in information about events that were on the news every day. The audience was united by shared concerns about political struggles to which they all were observers.

Creating inclusive communities

It takes more than abstract works or distant political debates to build inclusive communities and to transform, share and co-own spaces. To develop a sense of familiarity, shared use or ownership of diverse urban spaces, all involved parties need to feel welcome, comfortable and at home. While there is nothing wrong with large, formal interfaith lectures or events in 'neutral' facilities, the 'To Approach One Another' series illustrates that for local residents to really get to know each other, each other's religions, religious cultures, ethnic cultures, multi-cultural lifeworlds and respective spaces, it is paramount to create truly shared, egalitarian and frequent interactions in their own spaces and lifeworlds. In addition to dispersing information, events need to produce non-cognitive experiences. The moment when Mr Alwani and his audience laughed together about Nasreddin Hoça/Goha stories (a thirteenth-century folkloric comedian and Sufi whose stories and anecdotes are still well known across the Middle East) illustrates the power of shared emotions in the construction of feelings of togetherness and community. Not surprisingly, some participants noted that this was one of the best events of the series. Seminar-style spatial arrangements where participants face each other produce a sense of equality and allow participants to recognise each other on the street in their everyday lives. They make presenters part of the audience. Small gestures of welcome, like neatly arranged cookies, juice and water bottles in the mosque or the (*halal*) buffet in the Catholic community centre bespeak the desire to make everybody feel at home and create a sense of an open and shared space. Events without topics or formal discussions, like the women's breakfast, aim at good conversations and put participants in charge.

The series made a clear statement about gender issues without making much noise about it. Women simply participated. The role of Muslim women as presenters and also of significant topics, like the life of the Prophet, illustrates their public role and does more to end debates about 'oppressed Muslim women' than abstract lectures (Abu-Lughod, 2013). Women were part of the series in different capacities and this needed no further elaboration.

Visiting numerous local venues allowed participants to experience different spatialities and insert them into their own mental maps. This increased the nodes in their map of welcoming spaces and generated more inclusive communal geographies and inserted new spaces (the mosque, the book store) into existing spatialities and socialities. The series' goal was to approach another and to get to know each other and the respective other religion, its traditions and cultural context. Over the course of a year the series achieved more than informing local constituencies about each other. It created spaces of dialogue, memorable experiences, experiences of togetherness, and produced new or intensified existing networks between individuals and communities. The series introduced new modes and spaces of meeting, sharing and exchange.

About 1,200 individuals participated in the activities of the series run by the Evangelische Kirche Stuttgart (The Protestant Church Stuttgart). The remarkable effort of the church and mosque did not go unnoticed. Outside observers and journalists were among the participants. In late 2010, before the end of the series, the project was mentioned as an exemplary on the pages of the German Islam Conference (*Deutsche Islam Konferenz*, a government sponsored debate series). Shortly after the end of the series, the organisers were honoured with the Integration Prize of the regional council of the Protestant Church (*Evangelische Landeskirche*) (see www.stuttgart-evangelisch.de/ for the church's events and initiatives).

Concluding remarks

In the 'To Approach One Another' series, diverse women and men met in local venues, listened to presenters and music, shared food, attended a theatre, travelled to another city and participated in other activities. The organisers hoped to bring local residents together, teach them about each other, create new social and cultural networks and strengthen existing ties, and produce a new religious geography and a more inclusive sense of local ownership and belonging. Judging from the numbers of attendees, the lively conversations and evaluation by many visitors, and the prize and attention the series received beyond the neighbourhood, the series was a success.

The central success of the series is the resulting, even if minute, reworking of local religious and cultural geographies and the spatial, cultural and emotional grammar. To visit a new place is a first step in the extension of an individual's mental map. To feel welcome and comfortable in this space challenges or even undoes earlier misperceptions or resentments about this space and its owners. Seating arrangements or symbols of hospitality contribute to the creation of positive experiences and a sense of welcome, familiarity and community. The events in the mosque, its community's hospitality and experiences of casual conversations between Christians, Muslims and others signify significant moments in a German urban landscape where mosques are often treated with mistrust and resentment. In addition to showing that this prayer space can be used for mundane purposes and by non-Muslims, Mr Yilmaz' presentations also illustrated that a secular space (room at his work) can be turned into a prayer space.

The 'To Approach One Another' series illustrates that the minutiae of mundane, daily and small events, encounters and interactions transform urban spiritual and cultural geographies, and individual mental maps. Houses of worship and ordinary sites of public and civic encounters are starting points for the making and remaking of such landscape. These nodes, however, need to be linked with each other and other relevant spaces through the efforts of engaged individuals and communities. In the face of considerable Islamophobia and negative media images, this is a monumental task which can only be approached with persistent steps and engagements. Trust, friendship and community-making take time and effort. The organisers recognised the magnitude of the task and organised this ambitious series as a response.

The efforts of the organisers were not limited to the 33 events in 2010/11, but the church and mosque community continue their efforts. In emails usually starting with 'the To Approach One Another series continues', I receive occasional invitations from Mr Serdar. In November 2017 a Christian theologian spoke about 'The Bible in the Koran' and in October 2018 a Muslim theologian discussed the question of 'Religious Education and Theology: Completely Unnecessary for a Believer?' Invaluable relationships and networks are being carefully maintained and strengthened by more hard work, dedication and meaningful interactions.

References

Abu-Lughod, Lila (2013) *Do Muslim Women Need Saving?*, Cambridge, MA: Harvard University Press.

Azzara, Monique (2019) 'Grappling with the impermanence of place: a black Baptist congregation in south Los Angeles', *City and Society* 31(1): 77–93.

Baden-Württemberg, Staatsministerium (2005) *Muslime in Baden-Württemberg*. Stuttgart: Bericht für den Ministerrat.

Beaumont, Justin and Baker, Christopher (eds) (2011a) 'Postcolonialism and religion: new spaces of "belonging and becoming" in the postsecular city' in Justin Beaumont and Christopher Baker (eds) *Postsecular Cities*, London: Continuum, 33–49.

Beaumont, Justin and Baker, Christopher (eds) (2011b) *Postsecular Cities*, London: Continuum.

Becker, Jochen, Klingan, Katrin, Lanz, Katrin and Wildner, Katrin (eds) (2014) *Global Prayers: Contemporary Manifestations of the Religious in the City*, Zurich: Lars Müller.

Bowen, John (2007) *Why the French Don't Like the Headscarf*, Princeton, NJ: Princeton University Press.

Bowen, John (2010) *Can Islam Be French?*, Princeton, NJ: Princeton University Press.

Cloke, Paul and Beaumont, Justin (2013) 'Geographies of postsecular rapprochement in the city', *Progress in Human Geography* 37(1): 27–51.

Desplat, Patrick and Schulz, Dorothea (eds) (2014) *Prayer in the City: The Making of Muslim Sacred Places and Urban Life*, Bielefeld: Transcript Verlag.

Dwyer, Claire, Tse, Justin and Ley, David (2016) '"Highway to Heaven": the creation of a multicultural, religious landscape in suburban Richmond, British Columbia', *Social and Cultural Geography* 17(5): 667–693.

Fesenmyer, Leslie (forthcoming) 'Bring the Kingdom to the city: mission as place-making practice among Kenyan Pentecostals in London', *City and Society*.

Garbin, David (2013) 'The visibility and invisibility of migrant faith in the city: diaspora religion and the politics of emplacement of Afro-Christian churches', *Journal of Ethnic and Migration Studies* 39(5): 677–696.

Garbin, David and Strhan, Anna (eds) (2017a) *Religion and the Global City*, London: Bloomsbury.

Garbin, David and Strhan, Anna (eds) (2017b) 'Introduction: locating religion and the global city' in David Garbin and Anna Strhan (eds) *Religion and the Global City*, London: Bloomsbury, 1–24.

Habermas, Jürgen (2006) 'Religion in the public sphere', *European Journal of Philosophy* 14(1): 1–25.

Hervieu-Léger, Danièle (2002) 'Space and religion: new approaches to religious spatiality in modernity', *International Journal of Urban and Regional Research* 26(1): 99–105.
Kong, Lily (2001) 'Mapping, "new" geographies of religion: politics and poetics in modernity', *Progress in Human Geography* 25(2): 211–233.
Kong, Lily (2010) 'Global shifts, theoretical shifts: changing geographies of religion', *Progress in Human Geography* 34(6): 755–776.
Kuppinger, Petra (2011) 'Vibrant mosques: space, planning and informality in Germany', *Built Environment* 37(1): 78–91.
Kuppinger, Petra (2014) 'Mosques and minarets: conflict, participation and visibility in German cities', *Anthropological Quarterly* 87(3): 793–818.
Kuppinger, Petra (2015a) *Faithfully Urban: Pious Muslims in a German City*, New York: Berghahn Books.
Kuppinger, Petra (2015b) 'Pools, piety, and participation: a Muslim women's sports club and urban citizenship in Germany', *Journal of Muslim Minority Affairs* 15(2): 264–279.
Kuppinger, Petra (2017) 'At home in the multi-cultural city: Islam and religious place-making in Stuttgart, Germany' in David Garbin and Anna Strhan (eds) *Religion and the Global City*, London: Bloomsbury, 205–221.
Landeshauptstadt Stuttgart (ed.) (2013) *Datenkompass Stadtbezirke Stuttgart*, Stuttgart: Statistik und Informationmanagement.
Levitt, Peggy (2008) 'Religion as a path to civic engagement', *Ethnic and Racial Studies* 31(4): 766–791.
Livezey, Lowell (ed.) (2000) *Public Religion and the Urban: Faith in the City*, New York: New York University Press.
Orsi, Robert (1999) *Gods in the City*, Bloomington, IN: Indiana University Press.
Stepick, Alex, Rey, Terry and Mahler, Sarah J. (eds) (2009) *Churches and Charity in the Immigrant City*, New Brunswick, NJ: Rutgers University Press.
Tweed, Thomas (1997) *Our Lady of the Exile*, New York: Oxford University Press.
Werbner, Pnina (2002) *Imagined Diasporas among Manchester Muslims*, London: James Currey.

5 Praying in plain sight?

Women seeking privacy in the public realm in a Greek Orthodox Cypriot city

Anna Papadopoulou with Juliet Davis

Abstract

This chapter explores evolving gender relations, through the study of spaces of religious worship, within the historical core of the Cypriot city of Limassol. It identifies three main types of religious spaces. Two of these are associated with the Orthodox Church of Cyprus which has long been the dominant religion on the island. These two are distinguished primarily according to scale, the first being large, grandiose churches for public events, and the second comprising tiny churches and chapels typically set within the urban fabric, which are used more by women. The third is a type that has only recently emerged within the city, as a result of migration from South and Southeast Asia. Migrant groups are characterised by a varied range of denominations and religions and have appropriated a range of existing buildings including retail units, for worship. The ways in which these different spaces have been inhabited by women as manifest by the research is indicative of broader changes regarding the role of women in economic life within a patriarchal society in transition within an increasingly multicultural setting.

Key words: Greek Orthodoxy, Cyprus, religious spaces, domestic workers, immigration, privacy

Introduction

This chapter is concerned with exploring evolving gender relations through a study of spaces of religious worship within the historical core of the Cypriot city of Limassol, recognising these as examples of 'gendered spaces' as theorised by Daphne Spain (1992). With a metropolitan population of approximately 300,000 inhabitants, Limassol is the second largest city in Cyprus, following the island's capital, Nicosia. For a small city, it has a diverse and buoyant economy including financial and administrative services, tourism and a range of commercial activities. In terms of development, it is a city that bears the traces of the many waves of invasion and colonisation of Cyprus by foreign powers – from the Franks and Venetians in the Middle Ages, to the lengthy Ottoman occupation between 1571 and 1878, to the British from the late nineteenth century until 1960 (Kazamias, 2007; Hadjipaschalis et al., 2017). While this has undoubtedly shaped Cypriot society over the centuries, the majority of Cypriots today identify themselves as Greek-Cypriots whose native language is Greek (Kazamias, 2007). According to the Cyprus Population Review, as of 2018, 95 per cent of the population of the Republic of Cyprus adhere to Orthodox Christianity.

The Eastern Orthodox Church is the second largest branch of Christianity and it is dominant in southeastern Europe and Russia (Rohmann, 1999). It was first formed through a schism from Roman Catholicism which was long in the making but which finally occurred in 1054. Key contributors to this were differing perspectives on the status of the Holy Spirit within the Trinity, the supremacy of the pope and the organisation of the Church (Parry, 2010; Herrin, 1987). The Church of Cyprus is recognised as one of the oldest autocephalous churches in Eastern Christendom. Its history goes back to the first century when St Paul travelled to and preached on the island. The Church was well-established throughout Cyprus by the fourth century.

Limassol is one of a number of bishoprics within the Church. Reflecting the Church's historic importance and dominance, the city centre exhibits several churches that were constructed in the late nineteenth and early twentieth century. Their architecture generally typifies the historic church architecture of Cyprus which is strongly influenced by neo-Byzantine and neo-Gothic traditions (Andreou, 2006; Georgiou, 2018). Most were constructed on the foundations of older, smaller churches and monasteries (Andreou, 2006; Hadjipaschalis et al., 2017; Michaelidou, 2000). In contrast to these grandiose churches, such as the Agia Napa Cathedral and the church of Agia Triada (Figure 5.1), which were constructed to hold large congregations, there are also tiny churches and chapels which are often hidden within the urban fabric. These private expressions of faith, all of which date to the Ottoman occupation, would suggest a connection between the marginalisation of Christian communities within the public realm. From the scriptural emphasis on a male God, messiah and saviour-king, male apostles and evangelist saints to the division of and boundaries between gender-specific areas in church buildings, the Church is key to an understanding of the history of patriarchal society in Cyprus.

There is another type of religious space that has, in the past 20 years, surfaced within the evolving urban fabric of central Limassol. This is characterised by the appropriation of existing buildings including retail units. Such spaces can be found peppered between shops and reflect the influx of migrants into Cyprus from elsewhere in South and Southeast Asia, mainly the Philippines, Sri Lanka, Bangladesh and Vietnam, representing different Christian denominations – such as Evangelical Christianity, New Testament Church and Seventh-Day Adventist Church – and other religions, including Buddhism.

The research set out to explore how, first, these different religious spaces were historically designed to be used and inhabited by women and, second,

Figure 5.1 Agia Triada.
Source: Anna Papadopoulou, 2016.

how all three types are used today. Its aim has been to consider what this suggests more broadly about transitioning gender relations in Cypriot society. Several key findings arising from it are identified through the chapter. First, the two types of Orthodox churches reflect gender distinctions that study of their contemporary use shows to be no longer current. This changing use can be connected to the entry of women into the workforce in the 1970s and the increasing presence of women in the public realm, while at the same time ongoing association of women with devotional spaces in the city centre. In turn, some of the recent ad hoc, multi-denominational spaces are found to reflect a particularly female ethnic diversity related to the influx of domestic workers, a phenomenon also connected to the transforming role of Cypriot women in the economy.

The analysis and argument that follows is structured in four parts. In the first, we provide a general overview of the role of women in contemporary Greek society, drawing on studies on Limassol by sociologists and cultural historians. In the second, third and fourth, our analysis of the three types of religious spaces is set out (Figure 5.2). In the conclusions, we summarise the findings, particularly in terms of how the evolution and diversification of religious spaces is connected to the development and transformation of gender and class relations in Cypriot society (Figure 5.3).

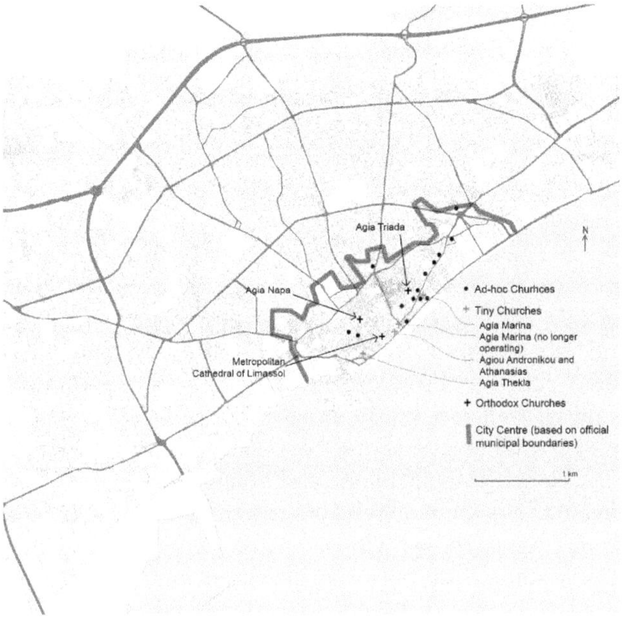

Figure 5.2 Map of Limassol, showing three types of religious spaces within the city centre.

Source: Anna Papadopoulou, 2019.

Figure 5.3 Women in church, circa 1940.

Source: Patticheion Municipal Museum, Historical Archives, Research Centre of Limassol (used with permission).

Women in Cyprus and Limassol

Research focusing on Cypriot women's role within the social structure conducted in the last 30 years suggests that Cypriot women have been in a transitional phase between a status of submission and liberation (Hatzipavlou, 2004; Mylona et al., 1982; Pyrgos, 1993). Traditionally, the place of a Greek-Cypriot woman was in the home. Christianity played an important part in designating women's place within social structure and, indeed, the Greek Orthodox Church is clear on women's necessity to marry and procreate.

However, since the 1970s, this has been changing, as elsewhere in the Western world, if perhaps more slowly. Women's entry into the workforce has been an important part of this, carrying implications for the geography of male and female activities and labour across the city. The role of women has been increasingly distributed between the home and the workplace, creating challenges that are not always well met by the urban form and layout of the city, manifesting a range of spatial and temporal boundaries (Papadopoulou, 2018; also see Andrew, 2000; Greed, 1994; Miranne and Young, 2000)

Social changes have had a bearing on religious practice and perceptions of the Greek Orthodox Church. Research by Mylona et al. on Cypriot women aimed to evaluate the degree of women's psycho-social liberation in the 1970s. It considered women under 24, women between 25 and 39 and women over 40. According to Mylona et al., one-third of younger people were prepared to express scepticism with regard to the dogmatic and ceremonial presence of the Greek Orthodox Church, though two-thirds remained quite religious. The women of the latter demographic attended church frequently and possessed a deep faith that God will provide solutions to their problems (Mylona et al., 1982).

According to further research, Cypriot women consider themselves socially progressive and independent and on the other hand, they have been found to be unaware of patent gender biases both socially and professionally that prevent them from being equal members of society (Hatzipavlou, 2004; Pyrgos, 1993). For instance, in a survey on Cypriot women's social profile conducted in the 1990s, a high percentage of women said they felt there was equality between genders. At the same time, they also confirmed that Cypriot society was very much still a patriarchal one, seemingly unaware of the contradiction created by these two positions (Hatzipavlou, 2004).

Since Cypriot women's entry into the workforce in the second half of the twentieth century, a new kind of women has emerged in Cypriot cities. These women, who come mostly from South and Southeast Asia, take on the domestic workload Cypriot women can no longer maintain because of their duties as members of the labour force. On Sundays, their day off, these women flood the city's most famous public areas, as they gather in public open spaces to socialise, attend religious gatherings and exchange goods and services. They inhabit the city's coastal promenade park for picnics, camp on large pavements in front of buildings to cut each other's hair and barter goods and pose in front of the city's monuments to take photos to send back home (Figure 5.4). Their presence in

Figure 5.4 Domestic workers on Sundays.
Source: Anna Papadopoulou, 2016.

Limassol's public realm on Sundays is not only proof of the city's growing social diversity, but also serves to activate public and interstitial spaces.

The active use of public spaces by foreign workers stands in contrast to the habits of native Limassolian women. As Papadopoulou's research shows, despite being participants in the labour market, Limassolian women's public presence continues to occur less in the world of streets, pavements, squares and parks (Papadopoulou, 2018). According to a 2016 Eurostat survey, Cyprus ranks fifth in Europe in car ownership per 1,000 citizens. In the case of Limassol, this can be partly attributed to an inefficient system of public transport, i.e. buses, and to an ever-increasing suburban sprawl that extends northwards from the city centre.

Orthodox churches

Over the centuries, large Orthodox cathedrals and churches have been constructed within Limassol's city centre, their plots typically ranging from $500\,m^2$ to $800\,m^2$. Many were constructed in the first quarter of the twentieth century on the sites of earlier churches (Andreou, 2006; Michaelidou, 2000). The churches' sheer scale and solidity allows them to dominate the surrounding built environment as they are approximately the size of an entire urban block and their domes can reach as high as 30 metres. The grounds surrounding most of the churches have assumed the character of a public gathering space that is occasionally used for religious events but can also function as an informal public space. Their importance in local culture is not only related to their monumental size, imposing architecture and meticulous iconography, but to their essence as symbols of ethnic pride, community unity and neighbourhood identity (Figure 5.3). The latter is evidenced by the fact that many municipalities in the older parts of Limassol, and indeed in every city in Cyprus, are named after the parish's church.

Beyond the established convention that assigns women and men to private and public urban domains respectively, spatial boundaries resulting in gender segregation are deeply embedded in architecture and within the built environment in Limassol, as in other cities in Cyprus. The spatiality of formal religious habits such as in Greek Orthodox churches reflects social relations and socio-spatial boundaries associated with them. The churches typically comprise three spaces, each revealing gender distinctions rooted within church life and social organisation: the sanctuary, which is accessible only to the priest, who is always a man; the nave, which is where the worshippers stand or sit; and the Narthex, which is the space that connects the nave to the outside world. Although this is no longer the case, historically, men would sit in the pews on the right-hand side of the priest and women on the left (Persianis, 2007). Placing male worshippers to the right of the priest further signifies man's supremacy to women. Since churches in the beginning of the twentieth century were unique public spaces where both men and women congregated, their separation through gender-specific seating areas became the prototype seating arrangement for other public buildings in Cyprus, such as theatres (Persianis, 2007).

While some argue that separating women from men inside the church space was a practice sanctioned and reinforced by religious patriarchies, it undoubtedly enabled men to maintain their position of privilege within Cypriot society. As a child, Anna (first author) remembers her grandmother explaining to her, her sister and her cousins that men and women sit separately in church to protect men from any unpleasant odours emanating from women attending church while menstruating. In fact, in certain parts of Cyprus, it was considered advisable until well into the last quarter of the twentieth century for women not to attend church at all while menstruating.

Although contemporary Cypriot society has become more secular over the years, the older generation of women – those born in the first half of the twentieth century – remain more loyal to religious worship than men (Papadopoulou, 2018). It was rare for women of this generation to have been employed, and as such, Sunday church was a way for them to escape from their daily routine of domesticity while socialising with other women from their parish. Weddings, christenings and funerals provided women with opportunities for affirmation of solidarity within one's family and social networks (Attalides, 1981). Although men also attend church, particularly on major religious celebrations, the majority of frequent churchgoers in Cyprus to this day are women, indicating a certain adherence to sanctioned gender roles. At the same time, the fact that women and men no longer sit in separate zones is reflective of a broader progress in social relations.

While older Cypriot women are at church on Sunday mornings, their husbands often meet at the local coffee shop. This was especially true during the first three-quarters of the twentieth century, when both men and women had less access to media and fewer possibilities for human contact through telephone or the Internet. Although housewives would socialise at church only once a week, men could visit the local coffee shop every day, where they would talk about work, politics and sports over a coffee and a game of backgammon. The coffee shops are informally called *traditional* coffee shops because their qualities are reminiscent of village coffee shops that played an integral part in the social life of Cypriot agrarian societies. In villages, these coffee shops were always centrally located and they were often adjacent to the village church.

Although the dimensions of traditional coffee shops vary, they generally comprise an indoor and an outdoor seating area and a kitchen large enough to accommodate a single worktable where the coffee was made and space for a fridge for cold drinks that would usually fit under the worktop. One particular spatial characteristic of the coffee shop that has been common in both its rural and urban expressions is the permeable treatment of the wall – usually the façade – that separates the indoor from the outdoor sitting area so that patrons can maximise their view of the public realm (Figure 5.5). This enabled the husbands and fathers of the women attending church, to directly observe all activity taking place on the grounds outside the church. While sipping their coffee, they could have a clear overview of who entered the church and with whom, what time services ended and decided if it was time to accompany their wives and daughters home. Traditional coffee shops generally open in the

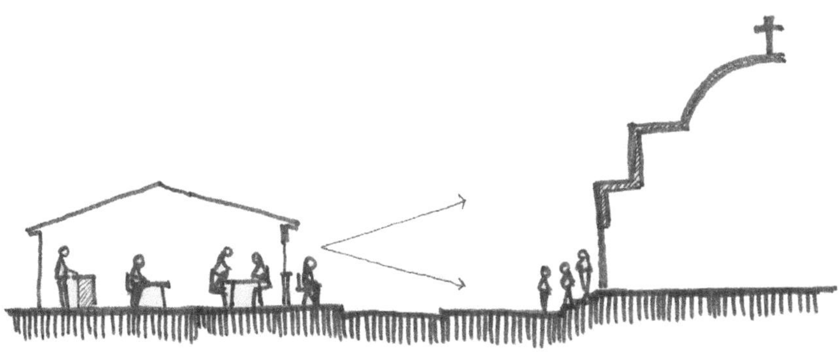

Figure 5.5 Proximity of churches and coffee shops in traditional village settings: the coffee shop on the left and a church on the right are adjacent to each other, and the arrows abstractly represent the (male) patrons' visual access to (mostly female) church goers.

Source: Anna Papadopoulou.

mornings and stay open until early afternoon because evenings are meant to be spent with the family at home.

This spatial connection that visually links the coffee shop to the church grounds does not figure as strongly in contemporary cities as it does in rural settings. In their urban format, traditional coffee shops are located in areas of mixed use, as close as possible to commercial pockets or other public buildings and uses in Limassol. Although the research supporting this chapter considered adjacencies between the location of traditional coffee shops and the churches, the results were not conclusive (Papadopoulou, 2018). This may reflect the fact that urban economic models and land-use values differ greatly from those in rural areas.

As some of these traditional coffee shops survive to this day in the centre of Limassol, the patrons remain exclusively men over the age of approximately 60. Younger generations increasingly do not subscribe to social segregation by gender, so the coffee shops today tend to be associated with an older demographic. In older times, the terms of women's exclusion from the coffee shops was clearer than today; although paradoxically, women would frequently be the proprietors of coffee shops, the institution of the traditional coffee shop defined these spaces as men-only zones (Figure 5.6). Women would never be seen sitting with the patrons. Although women's progressive emancipation has made the boundaries of gender segregation less stark, even in present-day traditional coffee shops, it is extremely rare for women to sit down and have a beverage there.

Figure 5.6 Woman serving coffee at a traditional coffee shop in Limassol, circa 1970.

Source: Patticheion Municipal Museum, Historical Archives, Research Centre of Limassol (used with permission).

This ambiguous socio-spatial space between the church door and the coffee shop's window pane that enabled men to exercise visual control on women attending church is sharply contrasted to a series of smaller churches hidden within the urban fabric. These tiny churches can be viewed as sites of resistance where women can seek privacy, away from the patriarchal stare, within the public realm.

Tiny churches

While attempting to unpack the significance of religion in urban gender politics by considering Limassolian worshippers' spiritual destinations, the research uncovered a different kind of church that sharply contrasted the cathedral-type churches that visually dominate Limassol's public realm. These are tiny churches, hidden within Limassol's urban fabric, with inconspicuous entrances that can easily escape the eye (Figure 5.7). Tucked away and concealed behind street façades, these hidden churches are considerably smaller in size, often comprise a single room and are almost always made of humbler building materials. Spirituality in these churches is experienced differently than in the larger churches. When looking up at the imposing dome of the latter, the worshipper is humbled by the might and power of the male God, whereas when entering one of these tiny churches, the worshipper can be inspired by the personal nature of religion and faith.

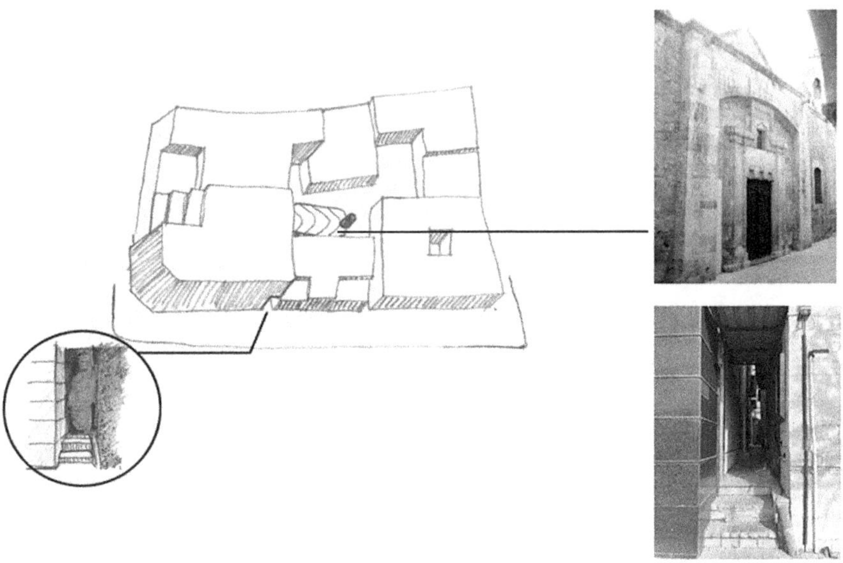

Figure 5.7 Religious spaces for local women: hidden church, Agiou Andronikou and Athanasias.

Source: Anna Papadopoulou, 2016.

These small churches suggest a blurring of the boundaries of public and private spaces on an urban and on an architectural level. Are these 'private' places of worship a response to women trying to carve out a place for themselves in the male-dominated public realm? Although research uncovered no universal evidence to support or dispute this, the existence of the tiny, hidden churches has afforded space for women to worship in privacy, reflecting their historic consignment to the private realm and the city's domestic sphere (Papadopoulou, 2018).

These hidden churches are not only a negotiated space that acts as a threshold between the public and the private, they also reflect connections between space, spirituality and patriarchy. These connections are revealed through traditions and tales passed down from generations that depict an intimacy between women and these smaller, more unassuming churches. Most cathedral-type churches in Limassol, and indeed in other cities, have a more gender-neutral character and lack the richness in feminine folklore and traditions that are encountered in the hidden churches.

One example is the church of Agiou Andronikou and Athanasias (Figure 5.8). It was said to have been a hiding point during the escape of a

Figure 5.8 Religious spaces for local women: hidden church, Agiou Andronikou and Athanasias, circa 2010.

Source: Patticheion Municipal Museum, Historical Archives, Research Centre of Limassol (used with permission).

woman known as Chiotissa – a woman from the island of Chios – and her companion, Anna, from the harem of a wealthy Ottoman (Michaelidou, 2000). According the story, the two women left the harem secretly at night, and ran to the church to hide. Inside the church, in a small storage area, they found two sets of men's clothes that had been left there for them by Chiotissa's brother. After they changed their clothes, they left their Anatolian, gold-embroidered dresses between the lit candles and the blessed oils. According to tradition, if a Limassolian woman wishes to accomplish a task, she leaves an item of clothing in that little storage room (Michaelidou, 2000). Near the well on the east side of the church, Chiotissa's brother and his friends were waiting while pretending to draw water. There they met the women and led them back to Chios (Michaelidou, 2000).

Another hidden church, Agia Marina (Figure 5.9), is located approximately 500 metres west of the church of Agiou Andronikou and Athanasias. It is a humbler structure, whose entrance was located behind a private house, approximately 15 metres behind a narrow road. Although the private house has recently been demolished, the entrance is still obscured and, as Agni Michaelidou (2000) writes, it is protected, by an old acacia tree. On the branches of the tree, mothers hang colourful cloths in the name of their sick children because St Marina is known as the patron saint of infants and children (Michaelidou, 2000). In addition to the cloths hanging from the tree, in recent

Figure 5.9 Religious spaces for local women: hidden church, Agia Marina, circa 2010.

Source: Pattícheion Municipal Museum, Historical Archives, Research Centre of Limassol (used with permission).

decades mothers of sick children have been commissioning local craftsmen to make wax replicas of a baby's body or of a child's ailing limb or organ and these replicas are then placed as a votive in the entrance of the church or hung on the branches of the old acacia tree.

Although the small churches considered in the research are attended by both sexes, there is a strong female demographic. Most of them have services on Sunday mornings and other religious holidays and remain open during the day for people – mostly women – who wish to walk in and light a candle or sit quietly in a dark, private space to pray, meditate or just be.

Ad hoc churches

Women from South and Southeast Asia come to Cyprus to work for approximately five to ten years so that they may earn enough money to ensure a better life for their families upon their return to their country. Frequently, domestic help will live with the family, especially if the female employer is a working mother, or she will come in once or twice a week and work for a few hours to clean and iron. If the domestic helper is not living with her employers, she will often share a small apartment with other women from her country. Domestic help that lives at the employer's home is responsible for a wider range of domestic tasks including childcare and care of the elderly, cleaning the house, doing laundry, preparing meals and doing some basic gardening. During their working day, these women remain enclosed in their employers' home, often isolated from human contact for many hours in a day.

The third type of space for religious worship the research uncovered is attended by women who are employed as domestic workers. The fact that the majority of worshippers are women is due to the fact that practically all domestic help are indeed women. These women adhere to a variety of religions such as Buddhism and several minor Christian denominations such as Evangelical Christianity, New Testament Church, Judah Ministry and Seventh-Day Adventist Church. Since there are no official statistics on ethnicities and their respective denominations, statements are based on personal observation.

These places of worship are located in Limassol's historic centre, which is also the city's commercial and administrative centre with occasional areas of mixed-use activity. These religious spaces are housed in shop spaces on commercial roads and are demarcated by a variety of signage. Some signs outside the shop are discreet, often little more than self-made banners or posters making the place of worship easy to overlook; other signage is more gaudy, colourful and welcoming. The glass panes of the shops are usually treated in a way to obscure the inside and provide some semblance of privacy for the worshippers (Figure 5.10). Other than chairs and tables for refreshments, the rooms inside are bare and unadorned, although some are equipped with a temporary stage and microphones. These spaces are approximately the same size as the hidden churches for Greek Orthodox women, making them considerably smaller than the cathedral-type churches that puncture that skyline of the city's historic centre.

Figure 5.10 Religious spaces for domestic workers.
Source: Anna Papadopoulou, 2016.

In spite of the presence of these worshipping houses' that blend with the assorted streetscape of miscellaneous commercial activity, these spaces are distinguished by the worshippers' singing, which can be heard outside. On Sundays, while Greek Orthodox churches announce morning services by the prominent sound of large bells, ad hoc churches emanate at around noon melodious sounds of singing and praying. The informal appearance of these places of worship encapsulates a series of metaphors for their users. Their improvisational and ephemeral nature speaks to an audience that is transitory and the intimate setting speaks to these women's search for comradeship among other women of their community. The socio-spatial boundary these women are called to overcome while living in Limassol is their daily physical isolation in the suburbs and other domestic neighbourhoods, and a feeling of nostalgia due to their lengthy absence from their home countries. Thus, these ad hoc churches manifest their inhabitants' space of negotiated privacy.

By meeting behind these shop fronts, in the city's public realm they are creating their own sites of resistance that mitigates these boundaries. At the same time, these women enjoy a sense of privacy in these spaces that is parallel to the intimacy of the Greek Orthodox hidden churches. Even the singing that is projected over loudspeakers acts as its own form of privacy and resistance; while the singing can be heard throughout the city centre, the origin of the singing is obscured by the density of the urban fabric, making them present, yet anonymous. These effectively employ temporal means to state their urban presence by religious songs that are less formal, yet more relatable to a society of growing secularism.

If domestic workers choose spaces of worship that are informal because of their desire for privacy, then how does this habit converge with their seemingly extroverted presence in the city's public spaces described earlier in the chapter? One interpretation is that what these women search for is a form of privacy in public that is conditional, a privacy that one experiences while being in an open space, tightly surrounded by one's countrywomen, engaging in activities reminiscent of one's home country, in spaces that are overlooked and bypassed by others. In a way, these women are hiding in plain sight, seeking anonymity in crowds, not unlike Susana Torre's account of the 'Mothers of Plaza de Mayo'.

In the mid 1970s, a small group of persistent women whose husbands and children had been secretly kidnapped by the Argentinian military dictatorship heroically demonstrated every Thursday at the end of the workday in order to call attention to their cause (Torre, 1996). The difference here between foreign domestic workers in Limassol and the Argentinian mothers is while the former employ crowds to create conditions of privacy, the latter group of women use crowds as a means of protection through anonymity. In the case of both groups of women, however, women's spatial appropriation is testament to the notion that public space is a physical sense not defined by buildings, but is elevated and engaged by social action (Papadopoulou, 2014; Torre, 1996).

Conclusion: what the eyes can't see

Women's evolving role within Cypriot social structures can be traced by the way they inhabit the public realm. Although in Greek Orthodoxy the priest and all Church members involved in delivering liturgy have been and still are male, women attending large cathedrals no longer have to adhere to rules that separated them from sitting among male worshippers. This indicates a certain degree of social progress in an otherwise stagnant status quo. However, by attending tiny churches, tucked away between buildings and obscured by the surrounding urban fabric instead of the cathedrals, Limassolian women are empowered by exercising their right to obscurity within the imposed formality of the public realm. They are negating the physical adjacencies between the church and the all-male traditional coffee shop encountered in rural settings and translated in urban environments, where men sitting at the coffee shop could observe women as they attended churches across the street. Women's intention of dismantling these socio-spatial boundaries is a subconscious remnant of rural life, because in contemporary cities such as Limassol the location of all-male coffee shops is no longer always across the street from church. At the same time, women living in Limassol who adhere to religions other than Orthodoxy exercise their own conditional privacy, by seeking less formal congregations, camouflaged within the city centre's commercial fronts, whose spatial characteristics reflect the temporary nature of their status in Cyprus.

Faith-based urban habits, although not always considered to their full potential by the professionals of the built environment, significantly contribute towards a more complete understanding of the political realities that govern urban form (Spain, 2001). From a secular perspective, the three types of religious spaces considered in this chapter – the cathedral-type church, the space behind the shop front and the small church hidden within the urban fabric – represent such notably formalistic opposites that defy a nexus between spatiality and spirituality. What links these women's religious habits together is an intrinsic desire and search for privacy within the city's male-dominated public realm, where tensions between genders persist on an architectural scale and are articulated as adjacencies and thresholds that create sites of resistance. Similarly, foreign women create occasions for spirituality in the heart of the male-dominated public realm in a way that dismantles conditions that bound them to their employers' house, away from the comfort of their fellow transient women workers.

Although Linda McDowell (1983) suggests that privacy is culturally specific, in the case of Limassol both local women and foreign domestic workers exhibit a parallel determination to gather in the public realm to pursue religious worship and a melange of other activities while preserving their right to privacy, suggesting that the need for privacy transcends race and income class. In the case of the women living in Limassol, privacy becomes a qualified, multi-layered condition determined by societal archetypes and political agency. As patriarchal order has confined women to the private realm of domesticity in order to maintain concentrations of power, even in domestic design there are instances of spatial programming that enable a theatre of observation, which in turn signify patriarchal control (Colomina, 1992). Thus, women's privacy cannot be ensured even in her designated 'place' within the private realm. In an attempt to safeguard their privacy from the architecture of control, Limassolian women have sought seclusion by establishing sites of resistance in the public arena such as the tiny churches, while foreign domestic workers seek privacy in worshipping behind shop fronts and in the anonymity of crowds.

However, the entire quest of pursuing privacy may be an act of contradiction that does not empower women but serves to reinforce existing boundaries. For that reason, bell hooks suggests that forms of domination can only be eliminated when the 'separation between domestic space/intimate space and the world outside ceases to exist' (Ainley, 1998). Privacy, according to hooks, is a 'screen for a profound narcissism', where mitigating it requires shamelessly spilling private acts, such as washing linen, into the public realm (Ainley, 1998). If Limassolian women are in fact transitioning between patriarchal submission and social liberation, and their desire for privacy in the public domain is being reconsidered in order to test a newfound autonomy by enabling faith worshipping to become more perfomative, their esoteric instincts spawned from generations of masculine violations of privacy will continue to inspire women to invent different avenues of resistance that will generate new architectural and urban form.

References

Ainley, R. (1998) 'Watching the detectors: control and the panopticon' in R. Ainley (ed.) *New Frontiers of Space, Bodies and Gender*, London and New York: Routledge, 88–100.

Andreou, T.A. (2006) **'Η Αρχιτεκτονική της Λεμεσού στα Χρόνια της Αγγλοκρατίας'** [Architecture of Limassol during the British rule], in ***Λεμεσός, Ταξίδι στους Χρόνους*** [Limassol, Journey through Time], Limassol: Municipality of Limassol.

Andrew, C. (2000) 'Resisting boundaries?' in K.B. Miranne and A.H. Young (eds) *Gendering the City: Women, Boundaries, and Visions of Urban Life*, Lanham, MD: Rowman & Littlefield, 157–168.

Attalides, M. (1981) *Social Change and Urbanization in Cyprus: A Study of Nicosia*, Nicosia: Zavallis Press.

Colomina, B. (1992) 'The split wall: domestic voyeurism' in B. Colomina (ed.) *Sexuality and Space*, New York: Princeton Architectural Press, 73–130.

Cyprus Population (2018, 1 October) Retrieved March 2019 from http://worldpopulationreview.com/countries/cyprus/.

Eurostat (2016) Statistics explained: Passenger cars in the EU, 2016. Data extracted in April 2018. Retrieved February 2019 from https://ec.europa.eu/eurostat/statistics-explained/index.php/Passenger_cars_in_the_EU.

Georgiou, C. (2018) ***Η Αρχιτεκτονική των Κυπρίων επι Αγγλοκρατίας*** *1878–1960* [Architecture of the Cypriots during the British Rule 1878–1960], Nicosia: Ekdosis En Typis.

Greed, C. (1994) *Women and Planning: Creating Gendered Realities*, New York and London: Routledge.

Hadjipaschalis, E., Lyssiotis, P. and Lyritsa, A. (eds) (2017) *Cyprus Has Always Been Europe*, Nicosia: Zavallis Litho.

Hatzipavlou, M. (2004) *Women in the Cypriot Communities*, Nicosia: PC Printways.

Herrin, J. (1987). *The Formation of Christendom*, London: Fontana Press.

Kazamias, G. (2007) **'Η Εισβολή και Κατοχή'** [Invasion and occupation] in A. Marangou, G. Giorgis, T.E. Sklavenites and C.Sp. Staikos (eds) ***Κύπρος, Από την Αρχαιότητα εως Σήμερα*** [Cyprus, from Antiquity until Today], Athens: Kontinos Publishings, 678–697.

McDowell, L. (1983) 'Towards an understanding of the gender division of urban space', *Environment and Planning D: Society and Space* 1: 59–72.

Michaelidou, A.M. (2000) ***Λεμεσός, η Παλιά Πόλη*** [Limassol, the Old City], Nicosia: Typografia Zavalli (2nd edition).

Miranne, K.B. and Young, A.H. (2000) 'Introduction' in *Gendering the City: Women, Boundaries, and Visions of Urban Life*, Lanham, MD: Rowman & Littlefield, 1–16.

Mylona, L., Paschalis, C., Kalava, E., Patsalidou, N. and Erotokritou, A. (1982) ***Η Κύπρια Γυναίκα, Έρευνα*** [The Cypriot Woman: A Research], Nicosia: Proodos.

Papadopoulou, A. (2014) 'Gendered spaces and participation in the public realm', *Spaces and Flows: An International Journal of Urban and ExtraUrban Studies* 4(2): 1–10.

Papadopoulou, A. (2018) *Her City: Spatializing Gender Relations in a Cypriot City*. PhD Thesis, Cardiff: Cardiff University. http://orca.cf.ac.uk/111256.

Parry, K. (ed.) (2010) *The Blackwell Companion to Eastern Christianity*, Malden, MA and Oxford: Blackwell Publishing.

Persianis, P.K. (2007) ***Πόλεις και Πολιτισμός: Ο Ρόλος των Κυπριακών Πόλεων στη Δημιουργία του Νεότερου Πολιτισμού*** [Cities and Civilisation: The Role of Cypriot Cities in the Formation of Recent Civilisation], Nicosia: Intercollege Press.

Pyrgos, M.A. (1993) *The Cypriot Woman at a Glance*, Nicosia: Pyrgos Public Relations.

Rohmann, C. (1999) A *World of Ideas: The Dictionary of Important Theories, Concepts, Beliefs, and Thinkers*, New York: Ballantine Books.

Spain, D. (1992) *Gendered Spaces*, Chapel Hill, NC: University of North Carolina Press.

Spain, D. (2001) *How the Women Saved the City*, Minneapolis, MN: University of Minnesota Press.

Torre, S. (1996) 'Claiming the public space: the mothers of Plaza de Mayo' in D. Agrest, P. Conway and L.K. Weisman (eds) *The Sex of Architecture*, New York: Harry N. Abrams, 241–250.

6 Pentecostal and personal perspectives

Gender, faith, urban space and town planning

Clara Greed

Abstract

In the first part of this chapter the different phases and aspects of Pentecostalism will be explained. While many traditional, white churches have experienced decline, one of the fastest growing sectors of the UK is that of Black Pentecostal congregations, especially in London because of immigration from African Christian countries, including Nigeria, The movement is also strong globally, especially in the developing world including the continents of Africa and South America. In the second part I will give a retrospective auto-ethnography (Maynard and Purvis, 1994; Saunders, 2017; Greed, 2019) on Pentecostalism and how it shaped my life, as I was a Pentecostal convert myself in my late teens. This personal account will illustrate how religion, gender and the spatial urban context and town planning have interacted in my own life. In Chapter 15 the spatial and social problems that Pentecostal congregations encounter when dealing with the town planning system in the UK will be illustrated.

Key words: Pentecostalism, Charismatic movement, mega churches, auto-ethnography, Holy Spirit, inner city

Introduction

In the first part of this chapter the different phases and aspects of Pentecostalism will be explained. While many traditional, white churches have experienced decline, one of the fastest growing sectors of the UK is that of Black Pentecostal congregations (Burgess and Van Der Maas, 2003: particularly 1080–1083 on the sociology of the Pentecostal movement; Lindhart, 2012). This is especially the case in London because of immigration from African Christian countries, including Nigeria, as the movement is strong globally, especially in Africa and South America. But white Pentecostal churches have existed too since the early twentieth century although nowadays many may be better termed 'Charismatic' churches as will be explained. In the second section I will give a retrospective auto-ethnography on Pentecostalism and how it shaped my life, as I myself was Pentecostal convert in my late teens. This personal account will illustrate how religion, gender and the spatial urban context, and town planning, came together and led me to being a town planner. In Chapter 15 the spatial, technical and legal problems that Pentecostal congregations encounter when dealing with the town planning system in the UK will presented.

The growth and nature of Pentecostalism

What is Pentecostalism?

Pentecostalism is not, strictly speaking, a denomination of Christianity, but a major modern movement within Christendom. Pentecostals may be distinguished from many other Christian denominations because of their doctrinal position and spiritual experiences. One of the more recent phases of Pentecostalism, namely the Charismatic movement, is in many respects very similar doctrinally to the traditional Pentecostals. In addition to the initial phase of 'being born again' that is becoming a Christian (John Ch. 3, Ver. 5 and Ch. 3, Ver. 16), both movements believe in the importance of 'the Second Blessing' that is the Baptism in the Holy Spirit (in addition to an initial personal salvation experience). This experience is separate from Baptism in water, which is also required of Pentecostal Christians and many other Non-Conformist groups. The name 'Pentecostal' was given to such churches because it was at Pentecost that the apostles and disciples first received the Holy Spirit and began speaking in tongues (see Acts Ch. 2, Ver. 1–47; and Acts Ch. 19, Ver. 1–6). Traditional Pentecostals believe that speaking in tongues is the only or chief evidence of the Baptism in the Spirit, whereas Charismatics allow for a range of gifts of the Spirt being seen as proof. Charismatics are often members of the historical denominations including the Roman Catholic and Anglican churches, and are generally more likely to be 'white and middle class'. The Charismatic movement in Catholicism is strongly linked to the reforms and new outlook introduced under Vatican II (that is the Second Vatican Council, 21st ecumenical council of the Roman Catholic Church (1962–65), which focused on spiritual renewal and reunion (Schrek, 2017)) and the influence of leaders such as Cardinal Suenens (Suenens, 1976). For example the current Anglican Archbishop of Canterbury, Justin Welby, has declared he speaks in tongues (Sherwood and Siddique, 2019), and that he aligns with the Charismatic 'Alpha movement' (Hunt, 2004), but he maintains a formal liturgical form of public worship. Both Pentecostals and Charismatics believe that miracles and the full range of the gifts, fruits and ministries of the Holy Spirit did not cease at the end of the New Testament period, but that they are still in operation today. This contrasts with the views of some Evangelical and Protestant groups who believe such manifestations ceased with the end of the New Testament. But the Roman Catholic Church has always believed in miracles, especially healing, albeit administered through the priesthood (Catholic Truth Society, 2016). (Evangelicals are those Christians who believe in the Evangel (Gospel), especially in the importance of receiving personal salvation through being born again (John Ch. 3, Ver. 16).)

All Pentecostal groups (including Charismatics) put an emphasis upon the importance of 'sanctification', which is manifest in how believers live, and their dedication to 'good works'. This is manifest, particularly in modern Black Pentecostal churches, where congregations provide a wide range of facilities and ministries for 'good works' tackling educational issues, poverty, catering, housing and employment needs, all part of what Katie McClymont describes as 'municipal spirituality' in Chapter 10. The wide range of ministries, caring activities, healing events, miracles and spiritual experiences offered by Pentecostal churches has contributed to their attractiveness and therefore growth. Requirements for new, large, multi-purpose church buildings have inevitably contributed to a collision with the secular town planning professions (see Chapters 14 and 15), who have limited experience or understanding of such churches' spatial needs (Greed, 2016). Black Pentecostals generally hold a much more politicised outlook than found in many white Evangelical churches, and therefore are likely to campaign against the planning decisions that they see as a racist planning system.

Historical development

The modern Pentecostal movement dates back to the 1906 outpouring of the Holy Spirit on a small Black congregation in Azusa Street, Los Angeles, from which it spread remarkably rapidly (Gee, 1967: 119–127; Burgess and Van Der Maas, 2003: 344–350). I visited the site of the original Azusa Street church in 2013 expecting to find at least a cross or monument at this historically highly significant site. All I found, with difficulty, was one small municipal plaque on a nearby wall. The site has been totally redeveloped as part of the Japanese quarter of the city. So spatially, there are no sacred 'places' in Pentecostalism. Churches are more concerned with moving on to new, larger premises than with setting up shrines to the past. But, every believer's body is the Temple of the Holy Spirit (1 Corinthians Ch. 6, Ver. 19).

The Pentecostal movement spread across the USA in the first part of the twentieth century. Significantly, both black and white assemblies developed. Gradually the movement spread to Europe, especially northern Protestant Europe, and small congregations began to develop in the UK. In terms of social class many were located in the poorer, industrial regions, particularly in the working-class areas of Northeast England, and in the coal-mining communities of South Wales in the Valleys (Kay, 2000). Two main 'white' denominations developed in the UK, the Assemblies of God (AOG) and the Elim Church, both of which were first formed in the USA. Paradoxically, the AOG is still a major white middle-class denomination in the USA, with many highly educated and politically active members. But both Pentecostals and Evangelical Christians seem to be much more right-wing and 'establishment' in the USA (Wallis, 2005; Boot, 2014). So one should always be wary about assuming a particular denomination is predominantly working class in character, or that denominations are the same in every country, in terms of class, ethnicity and education (Burgess and Van Der Maas, 2003). In the UK Elim may be seen as a little more like the modern Charismatic movement in being slightly more middle class and more flexible on its definitions of the Baptism of the Holy Spirit and 'quieter' in public worship.

In the UK, early on a few Pentecostal Bible colleges were established including the AOG Bible College at Kenley in South London (Gee, 1967), which we will return to anon. Both men and women trained at the college and, in theory, both could be pastors of churches, but often the husband was the main pastor and the wife became the support pastor and 'played the piano and sang' while the man did most of the preaching. Nowadays they are more likely to play the electric guitar and sing!

Meanwhile from the 1950s, a wave of immigration began, especially from the West Indies (Caribbean), to meet the demands of labour shortages in post-war Britain. But on arriving in England many immigrants found that the historic denominations that used to send the missionaries are now in decline themselves, and the churches were found to be unfriendly to immigrants. Many Black people felt unwelcome in the cold, formal British churches, and doctrinally they did not find a place to express their particular brand of Christian worship. So they began to set up their own churches, usually in any vacant hall they could find, as they established a foothold in their new country. So the first wave of Black Pentecostal Christians mainly came from previous overseas territories within the British Commonwealth, especially the Caribbean, whereas the second wave has come mainly from Africa (Muir, 2004).

Many immigrants (both Christian and Muslim) come from home countries that have never been through a secular phase, and that do not separate the public realm from private life, or personal belief from political activity. Therefore they are likely to be seen as out of step with the UK's predominantly secular culture and society, and are likely to be viewed with suspicion. Many Black Pentecostals in the UK are immigrants from countries where there is no welfare state safety net, or right to free health care, and may find they have limited rights in their new country too (Freeman, 2012). So they are more likely to develop a culture of self-reliance, entrepreneurialism, dependence on the extended family and belief in the power of prayer to intervene in their disadvantaged lives. Many such churches stress the 'Four Square Gospel' that is not limited to some future heavenly experience but touches 'Mind, Body, Soul and Spirit', that is, all aspects of human life on this earth. (Their faith is not limited to the four walls of the church.) Taking literally the sentence in the Lord's Prayer, 'Thy Kingdom come on earth as it is in Heaven', they believe that their material circumstances will change as an outworking of spiritual blessing. All this contributes to a dynamic growing church, thirsty for more buildings and space.

Gender considerations

While Pentecostal churches offer a non-traditional approach to Christianity, and are diverse ethnically, they are a mixed blessing in terms of gender (Greed,

2017). The question of whether women as well as men are entitled to experience and operate the full range of ministry, as inspired by the Holy Spirit, has been a vexed one. We might all be saved and equal in the sight of the living God, but gendered cultural attitudes still restrict the chances of women operating their Spirt-given gifts and ministries. Some Pentecostal and Evangelical churches fully accept women's ministry, and their right to lead, speak, pray and preach in the assembly 'in public', whereas others take the view that women should remain 'silent'.

This attitude originated, many centuries ago, from a sexist, poorly translated, decontextualised interpretation of, and overemphasis on, some New Testament verses. It is transmitted down over the years through all branches of the Church both Catholic and Protestant. For example, 1 Timothy Ch. 2, Ver. 11–15 forbids women to teach or exert authority over men; and 1 Corinthians Ch. 14, Ver. 34 tells women to keep silent in church. In contrast, other verses attest to women prophesying (Acts Ch. 2, Ver. 17), teaching, preaching and being involved in church governance (Romans Ch. 16, Ver. 1–16), stating there 'neither male nor female' in Christ (Galatians Ch. 3, Ver. 28) (Adams, 1973). St Peter declared (1 Peter Ch. 2, Ver. 5 and Ch. 2, Ver. 9) the 'priesthood of all believers'. So, in theory, anyone, male and female, black and white, and from any social class, can become a prophet, priest, leader, healer and preacher, within the more progressive Pentecostal churches. Black women, in particular, are more prominent in church ministry than in traditional denominations (Beckford, 2000: 49). But as we found in previous research, this does not mean that 'the outside world' accepts women church leaders. For example, colleagues at Planning Aid in London suggested that some ethnic minority faith group appellants felt they were not taken seriously by male planners, who were more used to dealing with white vicars (Onuoha and Greed, 2003: 86–90). The gender issue has certainly affected my life both as a Christian and as a town planner. I will now describe my own life experience to show how gender, religion, planning, urban space and class interact.

My life path, Pentecostalism, planning, social class and gender

I now want to explain quite why I am so interested in Pentecostalism and continue to do research on its relationship with town planning. To do so I need to give a short account of my own life story (Maynard and Purvis, 1994; Saunders, 2017; Greed, 2019). While religion and planning are key factors in this account, my class and education experiences have also shaped who I am and why I turned to religion. I was born in 1948 in South London, as an only child, with elderly parents. Initially we were quite well off, and I went to a nice little private school. Both Mum and Dad were well educated and quite socially aware and Mum described herself as a free thinker, whereas Dad seemed to be very anti-religion of any sort. My family was not religious and never went to church. So most Sundays in the 1950s we used to go for rides in the car down to the coast and around the countryside. Early on I was aware of what I would later realise were town planning issues, such as the growth of the numbers of motor cars, and increased suburbanisation, as my parents commented that 'one day there will be so many cars that there will be no countryside left'.

I knew nothing about God, and was keen to find out about Christianity. So I attended the Scripture Union group which was a Bible study group for the young pupils at school. At school the teacher who ran the Scripture Union group always told us to give our lives to Jesus, and I did so many times in my heart but never found out any more about it until many years later. The teacher always used to tell us we must go to church but I could not imagine going on my own and was not sure where to go locally. Looking out of the car window I often saw people going to church on Sunday, and I wanted to see what one was like inside, but I was told it was not for me.

But ill fortune subsequently was to befall us and we experienced a drop in social status, and became quite *déclassé*. My life changed in my early teens when, owing to major financial, business, health and personal problems, we lost our business and ended up in a house in an inner city area which had gained notoriety as 'The Criminal Area' in Terence Morris's late 1950s study of the area in question (Morris, 1958). Actually the area was quite mixed, socially and ethnically, and we were still owner-occupiers, but under threat of clearance. Meanwhile having moved to a state school, I passed the 11-plus exams but I ended up in the D stream of a girl's state grammar school. I think a lot of the decisions of streaming were based upon the area where pupils lived, and their father's occupation (none at the time), and I was judged accordingly. Many years later, having married a Christian solicitor, we considered taking a class action against Morris's employer, UCL (University College London), for defamation of character (of both the area and its residents). But by then 20 years had passed and in any case English law does not work well for 'class actions'. Meanwhile, in my teens Mum was very ill and had detached retinas and spent a lot of time in Moorfields Eye Hospital, whereas Dad took to drink and endlessly playing Tchaikovsky on the piano. So my parents had very little contact with the school. For me it was a matter of survival, trying to look after the house and my parents, and then get to school the next day exhausted. The teachers offered no help: they did not even seem to know who I was. I seriously considered leaving when I was 14 to care for Mum, as it was all too much (the school leaving age then went up to 15 about this time).

I found it very difficult at school, and I probably looked rather uncared for, and I was always very thin and worried looking. At home the property next door was turned into a half-way house by the council for people with problems. We frequently had our windows broken, dog faeces and lighted newspaper put through the front-door letter-box, and it was sometimes quite dangerous to go out. I was shot at in my bedroom because I had foolishly put the light on without closing the curtains. Dad tried to get the police involved but they seemed both aggressive and disinterested, and threatened to arrest Dad. This was a huge shock to my parents, as in their previous life they were used to the police being deferential and helpful. No wonder I am a nervy person.

The neighbours really seemed to resent us, perhaps because they thought we were 'posh' and wanted to take out their social grievances and class hatred on us. The whole area was very multi-ethnic and this was only the early 1960s. I found the local Indians and Jamaicans quite friendly and enlightened, whereas the white people next door were terrifying. At the top of the road by the railway station there was a 'red light' district and so it was always very important to be careful when passing through the streets up there on the way to the bus stop to get to school. Teachers obviously saw me as part of the great unwashed and used to come up to me in the corridor and say 'Who do you think you are? You're nothing' and 'People like you aren't fit to have children' and 'You're selfish to want to have children' and 'You always have too many children and expect others to look after them'. Because they thought I was working class they seemed to assume I was immoral too, although my upbringing had been very sheltered. At the time I had absolutely no class consciousness or social awareness and took it all very personally and thought there was something terribly wrong with me and cried myself to sleep every night.

It all came to a head at school when pupils had to choose which O levels (GCSEs) they were going to study and I was put in the class where pupils were not allowed to take O levels. It so happened that Mum was not in hospital at this time, and she was absolutely amazed about all this, presumably 'trusting' the school and not having a clue what was going on. She put on her best grey suit, sorted herself out, took her French school certificates and Sorbonne diplomas with her and went to see the headmistress (she had been educated in France). The very next day I was told I was allowed to take eight O levels, and the headmistress who had previously always treated me like dirt suddenly started fawning over me and asking me if everything was all right, and actually started smiling at me in a deferential, solicitous manner. As a result of all this I am very cynical about the neutrality of educational selection and streaming as clearly class, or perceived social class, counted a great deal more than intelligence (Payne, 1983).

I worked like mad and got eight O levels, five of which had very good grades, and went on to the Sixth Form where I studied three A levels including history. I was still seeking God and started attending the local parish church, and eventually asked Mum if I could be Christened and confirmed. This caused no end of trouble, and I had to take my life in my hands to do this. So I was very disappointed to find that most of the congregation did not even speak to me and they were very snobby. Reading A level history, I was fascinated by the Reformation and the writings of Martin Luther and the other Protestant reformers (especially as discussed by Bainton, 1962) and I read about the Counter-Reformation too. I was totally entranced by Luther's idea that salvation was based upon 'justification by faith' by simply believing and not by works. This was a revelation to me, as all my life I had been told I was rubbish and not good enough. I also studied Renaissance art history and I remember looking at a painting of the Pieta, where the dead Christ is cradled in the lap of the Virgin Mary and like a vision it became very real to me that Christ had died for me.

Fortuitously, about this time my best friend from the local council estate had been converted in a Pentecostal house meeting, and invited me to attend a meeting over there. Two lady deaconesses from the local Bible college, namely Kenley, as mentioned above, were running an outreach programme, and I felt the presence of Jesus and gave my life to Him. They told me I was born again. But I told them I was 'justified by faith' just like Luther said, and was unfamiliar with the term 'being saved'. In retrospect there are many questions and contradictions in all of this. I subsequently found out that Luther had written some terrible treatises on both women and Jews (Luther, 1556). But I will never deny this experience as it became the basis of my future life and I continue in my faith, although my experiences with 'churches' have often been problematic.

In addition to the house church a few of us used to go to the newly established Black Pentecostal church nearby which was held in a run-down hall 'at the back of the sorting office'. We also used to go to other South London Black churches in Brixton, and were warmly welcomed, probably because we were young poor-looking little White girls and they felt sorry for us. I remember one Black woman praying for three of us, asking God to put a ring of fire around us, and I can still see the vision in my mind's eye to this day, and I believe this protected me when I entered the hostile world of town planning.

After all these early experiences, life settled down a bit at home, and I went to art college at Bath Academy of Art in Corsham, which was an associate college of Bristol University, run on Bauhaus principles. I was directed into 'art' as many girls were then, as few girls went on to university. Only 5 per cent of school leavers went to university in the late 1960s and most young people had left school by the age of 16 (Landau, 1991 (which says 80 per cent of pupils left school by the age of 15 in 1960); Ingham, 1981). I did my best to keep my faith going, mainly reading Meister Eckhart, the medieval mystic, not understanding the differences between Catholic, Protestant and Pentecostal reading matter. One morning I was in the bathroom in our hall of residence and suddenly I started speaking in tongues, which was a bit of a shock. I had heard about this but was not sure what it was, and I created a queue of other students wanting to use the facilities. Writing to the two deaconesses who ran my 'home' house church back in South London, I found out that this was the Baptism in the Holy Spirit and I was, in fact, Pentecostal.

I spent a year at Corsham but was very unsettled. We did a project for our three-dimensional design studies on town planning, and were asked to look at the layout and road system of our home town. So I went up to the planning office and found out all about the redevelopment plans. I did not know until then that it was possible to study town planning as a degree or that it was a profession. But Mum had had many encounters with the planners, as enemies, over the years fighting proposals to demolish our property and to put an enormous motorway through our area. It was now late summer and I tried to apply via the old UCAS university clearing system for a full time university place to study town planning, but got nowhere. I went to the Croydon careers guidance office and told the man who interviewed me I wanted to study town planning (as I had by then investigated the options in great detail), and he laughed and said I was 'clutching at straws' and directed me to a secretarial course.

I was so worried about my future that we all prayed about it a great deal in the house church. Not long after I went to a meeting at Kenley Bible College where they were sending missionaries off to the Philippines. Someone stood up in the row in front of me and prophesied 'The Lord will make a way for you'. Lucky old missionaries I thought! Then a woman stood up and prophesied, 'This is for you, the Lord will make a way for you', and I was knocked out with relief.

A short time after that, I heard from UCAS that I had been accepted, without an interview, on a new town planning degree course at a major technological university. The year comprised only 16 students with four girls including me. I walked in and immediately one of the boys said 'What are you doing here? This course is not for people like you'. Then my tutor informed that he could always tell which students were going to succeed and which were going to fail, making it very clear I fell into the latter category. He subsequently often told me I wasn't interested in planning, which amazed me as back home we lived and breathed fighting the planners: which was a much better education!

I was living in two worlds. Church members in my new Pentecostal church near the university comprised a range of types of people with a number of older women and men, some of whom were always requesting prayer for ill health, poor young families, manual workers, and quite a varied ethnic mix too. In contrast the planning course seemed to consist of young, male White boys who were never ill and who seemed totally unaware of the lives of ordinary people. The planning lecturers were obsessed with applying modern computer technology to planning, developing mathematical models to enable yet more motor cars and motorways to invade our cities. The whole emphasis seemed quite immature and adolescent in mentality, as both lecturers and students frequently proclaimed that 'women are stupid' and 'children are stupid' and 'housewives are always complaining'. One day, when in a seminar, the male lecturer was explaining to the male students that, with the introduction of computers, 'in the future nobody will have to work more than three hours a day and they can spend the rest of the time playing football'. Being the only female present I chirped up, 'But how can that be? A woman's work is never done?', with which the male lecturer turned on me and said 'Don't be stupid, we're not talking about that'. So what were we talking about? Every time I tried to raise the needs of women, I would be told that I should be concerned with 'the Blacks' (sic) in South Africa, and with 'the working class', neither of which group apparently included women. This was the beginnings of my feminist awareness.

After three years of sheer hell at university, I then went for the 'year out' to get practical experience being assigned to the local city planning office. Every day I experienced extreme sexual harassment (arguably assault) and also they would not let me go 'upstairs' to where the planning policy meetings took place. The whole experience shattered and disillusioned me, as there I was, ready to re-plan the world, and they would not even accept me as a planner, or respect me as a person. Similar experiences were recounted by other young women planners I met later on in Bristol. At the end of the year the planning school told me I was not fit to be a planner or to come back onto the final year of the course. So we resorted again to all-night prayer meetings back home, with much entreaty. Eventually I was allowed back and continued with the final year and then after much difficulty and many interviews got a job at a large rural county planning department. This was heaven in comparison with my previous experience, and they seemed decent types.

But one day I was told that a senior planning officer wanted to see me about my work, which was on public participation in relation to the new County Development Plan. He told me to sit down and then simply said to me 'You smell', and that he had had complaints about this. But nobody had said anything, and although by then I was a lodger in a Catholic Charismatic family which had eight children, I tried very hard to wash every day. I was quite used to being called 'filth' and 'dirt' when I was at school in the D stream but I had not expected this now I was a qualified town planner. (But I did sometimes have heavy periods.)

After this I gradually got into lecturing, mainly to chartered surveying students, and I was glad to get out of the planning office. I was quite anti-feminism as I could not relate to what I saw as rich, middle-class suburban American women, who had everything, moaning all the time. I had totally false gender consciousness as well as misplaced class consciousness. But gradually I moved towards feminism, firstly by reading about liberation theology, and thus into a more moderate English version of feminism which was more related to ordinary women's lives. I was now in my early thirties and wondered about doing a PhD. One of the early members of the 'women and planning' movement in London suggested that I research the position of women in the surveying profession with reference to both surveying education and professional practice, as surveyors have such an influence on property development and the nature of our cities. I started in 1985 and completed my PhD in 1990 under the supervision of the famous urban feminist geographer and economist, Linda McDowell (McDowell, 1983) and have never stopped writing, speaking and researching since.

By then I had become somewhat disillusioned about church and which one to attend. I had met my husband at Kenley Bible College when we both attended a short course for youth leaders on drug addiction, and he turned out to be a solicitor, law lecturer and lay preacher. We used to attend an Evangelical Anglican church for several years in Bristol. But what with doing my PhD and as well as working full time and looking after my mother after she had a stroke and came to live with us, we gradually wound down our church attendance. I was always looking for a church like the wonderful house church I was converted in, but I never found that special place again, and I roamed the spiritual wilderness. When I first came to Bristol I was pleased to find that there were Black West Indian Pentecostal churches in St Pauls. I went along to one all eager to join, only to be stopped at the door by a lady usher who looked me up and down and said 'We always put sociologists in the back row'! Perhaps it was a prophecy but at the time I did not know any sociology and I was extremely hurt.

So for years I have not been particularly a church person, but then I never really got the hang of how to be a Christian, as so much of it seemed to be cultural and one needed to be born in a Christian family to know the unwritten rules. After my husband's death from cancer in 2010 I went to a little old-fashioned (non-Pentecostal) gospel hall where the people were very supportive, as they were 'full of good works' experts in hospital and care home visiting. However, women are not allowed to speak or preach in the assembly, but when John died I just wanted to sit there in the silence. All my life, I have always had problems with this issue and remember several years ago when I first moved to the suburbs of Bristol that John and I went to a small pioneer Evangelical church for a while. John was always keen to preach but one day he did not feel well and asked if I would do so instead, as I was quite keen on preaching and teaching too. I had hardly got through the first paragraph of my sermon when I saw a kerfuffle happening at the back of the church, as the wife of the senior elder had gone out to find him and tell him what was happening. I was told to stop preaching and I never went there again. In contrast I have found a great welcome within academic and professional town planning circles and nobody (with some unfortunate exceptions) tells me I am not allowed to lecture or 'preach planning'.

On seeking Christian fellowship and a church to attend in recent years, I have found the whole credoscape (church landscape) has totally changed. When I was a student nearly 50 years ago now, it seemed everyone in the university town went to church, and chapels and church buildings were everywhere. When I go back to the university town now I find so many of these chapels have turned into restaurants, apartments or even mosques and temples. All those thousands of hymn-singing, Bible-believing Christians seemed to have vanished as if they never existed. But in South London, I find the opposite: new churches have sprung up, but the congregation is more likely to be Black than White, and the denominations are likely to be new and dynamic, while the historic White churches are still in decline. Back in my home town in the suburbs on the edge of Bristol, it seems that different churches cater for different age groups, perhaps reflecting the history and growth of the area itself. I have a choice of the Brethren gospel hall, elderly, few in number but very kind; or the Methodists and Anglicans, which seem also to be aging but with some influx of new families, and decidedly trendy women vicars, which never existed in the past. Or I can swap the quietness of the older churches for the rather noisy new Pentecostal church which has started up in our small town.

I prayed for years for a Pentecostal church in my small town, but now it has come I find it is nothing like the ones I knew in my youth. The church is undoubtedly thriving with more children than grown-ups in the congregation and an emphasis on attracting 'the youth'. Services are quite informal and everyone wears casual clothing (no suits and no hats for the women). There is a very loud pop group leading the service with electric guitars and drums, playing 'music' at high decibel levels. The result is that older people, like me, tend to sit at the back to protect their hearing. God is not dead, but God is not deaf either! (cf. Robinson, 1963, who declared 'God is dead'). There are no hymns, and no noticeable sermons, but lots of excitement. I miss the quiet meditative meetings of my house church long ago, where one could actually hear the words if someone decided to speak in tongues, pray, give a message or sing in the Spirit. But Christian friends tell me I am old and I must move with the times, and it is so important to attract young people to ensure the churches' future. This seems to be a bit of an insult to thoughtful, reflective young people who may prefer a bit of plainsong and quietness. While I have been a disaster as a church-going Christian, I still find both serendipity and blessing on my work, as town planning is still my spiritual vocation (compare with Chapter 3 where Inés Sánchez de Madariaga discusses Edith Stein's concept of secular professional vocations). So I continue to research the relationships between religion, gender, cities and planning. Indeed I find that nowadays one can even study 'Pentecostalism' as a serious academic pursuit at high-level universities such as SOAS and Heidelberg (Glopent, 2019–). How things change with time.

References

Adams, M. Queenie (1985) *Neither Male nor Female: A Study of the Scriptures*, Ilfracombe: Stockwell.

Bainton, Roland (1962 [1950]) *Here I Stand: A Life of Martin Luther*, New York: New American Library.

Beckford, R. (2000) *Dread and Pentecostal: A Political Theology for the Black Church in Britain*, London: SPCK.

Boot, J. (2014) *The Mission of God: A Manifesto of Hope*, Toronto: Freedom Press; St Catharine's, Ontario: Ezra Institute.

Burgess, S.M. and Van Der Maas, E.M. (2003) *The New International Dictionary of Pentecostal, Charismatic Movements*, Grand Rapids, MI, Zondervan.

Catholic Truth Society (2016) 'Sacrament of healing', *Catechism of the Catholic Church*, p. 325, para. 1421.

Freeman, D. (2012) *Pentecostalism and Development: Churches, NGOs and Social Change in Africa*, Jerusalem: Van Leer Institute.

Gee, D. (1967) *Wind and Flame: The Pentecostal Movement*, Croydon: Heath Press.

Glopent (2019–) The European Research Network on Global Pentecostalism, SOAS, University of London. see www.glopent.net.

Greed, C. (2016) *Christianity and Planning: Pentecostal Churches*, Occasional Paper 1, Faculty of Environment and Technology, University of the West of England, Bristol.

Greed, C. (2017) 'Sacred spaces' in S. Hawthorne (ed.) *Gender: God*, New York: Macmillan, 315–333.

Greed, C. (2019) 'My personal journey into planning', Paper presented at *Women and Planning Conference*, Leeds Becket University, School of Planning, 21 May.

Hunt, S. (2004) *The Alpha Initiative: Evangelism in the Post-Christian Era*, Farnham: Ashgate.

Ingham, Mary (1981) *Now We Are Thirty: Women of the Breakthrough Generation*, London: Methuen.

Kay, W. (2000) *Pentecostals in Britain*, Carlisle: Paternoster.

Landau, Cécile (1991) *Growing Up in the Sixties*, London: Macdonald Optima.

Lindhart, M. (2012) *Power in Powerlessness: A Study of Pentecostal Life Worlds in Urban Chile*, New York: Brill.

Luther, Martin (2015 [1556]) *Table Talk*. Edited by William Hazlitt. London: Beloved Press.

Maynard, M. and Purvis, J. (eds) (1994) *Researching Women's Lives from a Feminist Perspective*, London: Taylor & Francis.

McDowell, L. (1983) 'Towards an understanding of the gender division of urban space', *Environment and Planning D* 1: 59–72.

Morris, T. (1958) *The Criminal Area*, London: Routledge and Kegan Paul.

Muir, D. (2004) *Black Theology, Pentecostalism, and Racial Struggles in the Church of God*. PhD Thesis, London: King's College, University of London.

Onuoha, C. and Greed, C. (2003) *Racial Discrimination in Local Planning Authority Development Control Procedures in London Boroughs*, Occasional Paper 15, Faculty of the Built Environment, University of the West of England, Bristol.

Payne, Irene (1983) 'A working class girl in a grammar school' in June Purvis and Margaret M. Hales (eds) *Achievement and Inequality in Education*, London: Routledge in association with the Open University.

Robinson, J. (1963) *Honest to God*, London: SCM.

Saunders, P. (2017) *Croydon Boy: Growing Up in Post-War Britain*, London: Lulu.

Schrek, A. (2017) *A Mighty Current of Grace: The Story of the Catholic Charismatic Renewal*, New York: The Word Among Us Press.

Sherwood, H. and Siddique, H. (2019) 'I pray in tongues daily and sometimes hear prophecies, says the Archbishop of Canterbury', *The Guardian*, 21 January, p. 3.

Suenens, L.-J. (1976) *Ways of the Spirit: The Spirituality of Cardinal Suenens*, London: Seabury Press.

Wallis, J. (2005) *God's Politics: Why the American Right Got It Wrong and the Left Doesn't Get It*, Oxford: Lion.

Part III

Women, cities and Islam

7 Islamic and Christian cities during the medieval period

Iberia and North Africa

Carmen Hass-Klau

Abstract

The chapter compares and contrasts Christian and Muslim cities with particular reference to historical examples in Spain and North Africa. It provides the urban spatial characteristics of the Muslim City rather than centring on gender issues. The chapter shows how similar the urban organisation in these two cultures was in some respects. But some major differences can also be detected. These differences were regarded negatively by Western authors, for instance a lack of local government and organisation. This demonstrates the absence of awareness among Western writers regarding Muslim urban structure and the history of Muslim cities. Though both the Christian and the Muslim towns were to a large extent planned, the difference in the Muslim street layout was based on the freedom of the residents to create their own neighbourhoods. Muslim cities had much higher population densities than Christian cities. This was only partly overcome by migration into the suburbs. Western observers could not recognise the original street layout as over the centuries it had become too complex, hence they devalued Islamic cities. Hardly anybody saw the immense economic and cultural potential, which was created within Muslim cities by allowing people of different religions and ethnicities to live in separate neighbourhoods.

Key words: Christian cities, Muslim cities, Islam, Medieval, Spain, Morocco, local government, neighbourhoods, street layout, traffic

Background

Anybody who wants to read or write about the history of urban planning and transport in Spain and Portugal will inevitably get involved with issues of Islam and Christianity before getting engaged with cities. Nowhere in Europe was the confrontation between the two religions closer, fiercer and more inevitable than on the Iberian Peninsula, that is modern-day Spain and Portugal.

The process began with the arrival of a mixture of Arabs and mostly Muslim Berbers from North Africa at the beginning of the eighth century (711AD). In only seven years they conquered nearly all of Spain and Portugal and called it al-Andalus (Figure 7.1). They entered a country where the population lived within late Roman and early medieval structures. An Arabic dynasty (the Umayyads) arrived on the peninsula in 755AD, having lost its power in Damascus. The Umayyads created a new Caliphate in 926AD and Córdoba became the capital of al-Andalus; it developed into the largest city in Europe. It is estimated that the population was between 100,000 and 450,000 by the end of the tenth century. It became the main centre of medicine, mathematics, philosophy, astronomy and other intellectual professions. Its art, music and poetry were especially sophisticated and appreciated in many parts of the Islamic world. With the fall of the Caliphate in 1031AD, al-Andalus was divided into small *Taifas* (kingdoms).

The best land was taken by the Arabs whereas the Berbers inhabited the north and the northeast. A revolt of the Berbers in 740 against the ruling class

Figure 7.1 Map of Spain, circa 1100.

Source: Drawing by Helge Hillnhütter.

Note
Marked line: northern border of the Caliphate.

had the consequence that most Berbers moved to the south or to Morocco (O'Callaghan, 1975: 99). Some of the indigenous Christian population of the peninsula lived in the mountainous areas in the north and were known for being rebellious. When the Berbers left the northern region, this enabled the Christians under Alfonso I (693–757) to establish his kingdom of Asturias. The Christians had a first small but successful battle against the Muslims (in Covadonga) and this was seen as the beginning of the 'Reconquista'. It is a kind of irony of history that exactly the inhabitants of the northern regions, who had resisted most effectively the Visigoths, the Romans and everybody before them and were the least Christian, would become the champions of Christianity. The Reconquista was a kind of marketing campaign and it would go on for centuries. It would later be connected and mixed up with the Crusades.

Unexpectedly, the grave of St James was found in Compostela during the rule of Alfonso II (ca 760–842), which would bring money and political prestige for over a thousand years to this region. That the whole story of St James was totally irrational did not undermine its power. In fact, Compostela competed with Rome and Jerusalem in terms of its importance for pilgrims (Pöppel, 2014).

Weaknesses during the Muslim rule would increase the political and military power of Asturias and with the creation of other kingdoms in the north, the influence of the Christians grew. By far the most active kingdom to fight against the Muslims was Castile.

The Castilian armies conquered very early on, one of the major intellectual centres of the Islamic world, the city of Toledo, in 1085 (Watt and Cachia, 1996: 93). This was only possible because of the political fragmentation into smaller kingdoms, which had no problems fighting each other, quite often in alliance with Christian armies (Kennedy, 2014: 124–134). Only the military support from Berber dynasties from Morocco (the Almoravids, later the Almohads and after them the Marinids) slowed down the Christian conquest between the eleventh and the early fourteenth century. But Ferdinand III (1199–1252) of Castile was able to capture large parts of al-Andalus, including the two main cities: Córdoba (1236) and Seville (1248). After the fall of Granada in 1492, Spain became the most powerful and richest country in Europe.

This summary of the political events gives an idea of the closeness of the political and religious struggle between the Muslims and the Christians within the peninsula, which also included Morocco over nearly 300 years. Under the deception of a religious war between Islam and Christianity, which some of the main players must have believed in, the Iberian Peninsula became slowly part of the Christian world. Before and during this military conflict, the two worlds, which had been created since the eighth century, coexisted. This was expressed among other things in the way cities were built (Figure 7.2). But what was an Islamic city and how was it different to its Christian counterpart? Let's begin

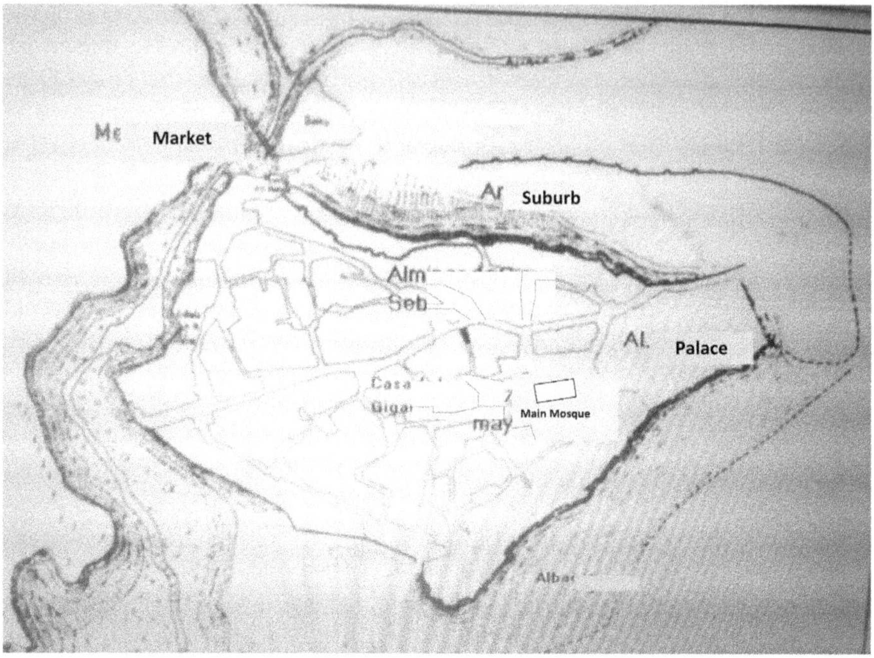

Figure 7.2 Typical Muslim town in al-Andalus; only the main streets are showing.
Source: Graham Crampton.

with one of the most controversial phrases in the literature of urban planning: the Islamic city.

The Islamic city: can it be defined?

The books and articles written on the 'Islamic City' are overwhelmingly numerous, complex and controversial, covering Muslim cities in many different countries. The early publications, which started in the late 1920s and continued up to the 1960s, studied especially North African towns (W. Marçais, 1928; Sauvaget, 1941; G. Marçais, 1940; Brunschvig, 1947; L'Tourneau, 1949; Grunebaum, 1955; De Planhol, 1968 etc.). A second wave of authors entered the debate, mostly in the 1980s, 1990s and more recently; they had a more critical approach to the earlier publications, only to mention a few, such as Lapidus (1973), Abu-Lughod (1987), Alsayyad (1996), Wirth (2000) and Petruccioli (2002). It is argued that the authors of the earlier period saw the Islamic city from the viewpoint of Western superiority. European planners, who had been interested in Islam, were of the opinion that their towns were unplanned, chaotic and the streets were labyrinthine.

Most authors emphasised that Islam was a religion that favoured an urban environment and there is more or less agreement that Islam is closely connected with city development in general. There were other opinions as well. Petruccioli (2002) mentioned De Planhol (1968) who believed that the nomadic origins of the Arabs led to a limited urban sense and the inability to form a coherent physical and social structure. He saw the nomadic element as persistent and continuously challenging society, which itself is seen as a unitarian body (Petruccioli, 2002: 202). The Arabian historian, Jaldūn or Khadun, from the fourteenth century believed (and De Planhol very likely copied him) that the problem of the Islamic city was the traditional nomadic lifestyle of the inhabitants (Balbás, 1970: 41).

Some explanatory comments about Islam are necessary here. The Prophet Mohammed came from a trading town and was a trader himself (Haleem, 2016: xi). The leading class of Arabs who conquered North Africa and al-Andalus were no nomads, and although large parts of the army consisted of Berbers of whom some were nomads and others were not, it was the ruling class who decided where to settle, where to found towns and how they were to be designed.

Grunebaum's main publication, *Medieval Islam*, first published in 1946, described that the Muslim Arabs, on their way towards creating an empire, absorbed the Roman traditions, laws, administration, finance and even their philosophy. It was from the Greeks and Romans that the Arabs learned abstract thinking. Islam had little historical tradition when the believers started their conquest (1953 [1946]: 3). His view on the cultural past of the conquering Arabs was very much in contrast to Lapidus (1973), who wrote:

> [T]he pre-Islamic past was the foundation of Muslim city experience. From the past came a highly differentiated society, stratified by classes, divided into subcommunities, and dominated by a hierarchical form of social organisation ... From the past came also the complex techniques of industries and crafts, norms of commerce, and market institutions.
>
> (1973: 52)

But the most negative and devastating view about the Islamic city in general came from the French historian Jean Sauvaget, who wrote:

> One can attribute to it nothing but the dislocation of the urban centre ... the work of Islam is essentially negative ... and the city is an inconsistent and inorganic assembly of districts ... and is nothing so much as the negation of urban order ... The Islamic city is no longer considered as an entity, as a complex and living being in itself, it is no longer anything but an assemblage of individuals and their contradictory interests.
>
> (Sauvaget, 1941: 247–248, quoted in Falahat, 2014: 13)

A whole chapter could easily be written about these different opinions on the background, the main ideologies and the physical ingredients of a Muslim city, considering that such cities are widely spread from the northern half of Africa, to the Middle East, Turkey, the Arabian Peninsula, Persia, Pakistan, Afghanistan and Indonesia, only to mention the most important countries. Therefore, one is confronted with large cultural differences and an extremely wide diversity when discussing the urban structure in general terms. These cities have very different characteristics in North Africa or in historic al-Andalus compared to the equivalent in the Near or Far East. One can conclude that general definitions are extremely difficult if they are supposed to be meaningful. But, in contrast, if one looks up the phrase 'Christian City' on the Internet or in the literature, it does not lead to a similar discussion. In order to simplify matters, a table (Table 7.1) with the main characteristics of the two types of cities is provided here and further discussed.

At first glance, there seemed to be many similarities between the Muslim and the Christian city, at least during the medieval period. However, variations are likewise visible:

a the local government structure within the Christian city
b the difference in the street layouts between the two types of cities
c the importance of the neighbourhoods in the Muslim city
d contrasts in the construction of private space.

Local government structure in the Christian city

One of Grunebaum's main criticisms was that the Islamic, in contrast to the Western city (he actual used a Greek city), was not an autonomous community (or a free city) and lacked a local government (1955: 139). Grunebaum, who was by origin Austrian and a historian, must have been well aware of Max Weber, who wrote an uncompleted essay titled 'Die Stadt' between 1913 and 1914 (Nippel, 2000: 103). Weber tried to prove that outside the occident, no development of urban local government took place but that local government was a precondition of capitalism and the modern state (Nippel, 2000: 111). He does not explain what he meant by capitalism and the modern state but if he meant making money and profits then he was certainly wrong. Although one can assume that with the increase in local government power, economic growth may follow, it was not automatic. In the case of the Western European towns this was mostly true from the twelfth to the fifteenth century. Yet many of the Muslim cities, despite the lack of self-rule, were economically extremely successful, and Islamic countries were a constant economic and political threat to a number of Christian societies, including Austria in 1529 (the first attempt to conquer Vienna). During the early sixteenth century, the Ottoman Empire (1299–1922) had become the main political power and had replaced Islamic Spain and was even threatening Morocco. In terms of economic success, the Muslims had beaten Venice, which was famous for its local government, in 1573 and controlled trade in the eastern Mediterranean. There are plenty of other comments one could

Table 7.1 Differences and similarities between the Muslim and the Christian city

The medieval Muslim town of al-Andalus or Morocco and the Christian city were characterised by an impressive town wall and had the following elements:

Muslim city	*Christian city*
Central power: The Caliph, or similar leader would occupy a palace, which was mostly separated from the main town and was located at the edge of a town. The size of the enclosure of the Muslim ruler could be enormous. It was so large that it was a town on its own with all the necessary skills and buildings, therefore often called the 'royal city'. It seems that all towns of importance had a quarter where the head of the country or region could reside. There was no town government but a religious codex on how people had to behave and controlling organisations over trade and laws. Self-rule of the town was an exception but it did exist.	Christian cities originally also had a central controlling power (king, dukes, lords) or the church (abbots, bishops). The castle would be at the edge of the town at the highest location, often quite similar to the Muslim palace. This could also be the case for monasteries. But from the late twelfth and early thirteenth centuries onwards, the control of the town/city was taken over by citizens. The typical medieval town was independent and had its own government, for instance the free (or imperial) cities of the Holy Roman Empire. Some of them had political power extending far beyond the city borders, and their own army. A number of these free cities became very powerful and would be part of a kind of 'club', which would represent their interests, such as the Hanse or the North Italian Leagues. The North Italian Leagues had their own army.
Street network: The Muslim towns normally had three types of streets. The main streets, which would run in an unobvious way through the town, the border streets for the neighbourhoods and the cul-de-sacs for the residential areas. Quite often it was not clear which ones were the main and which ones were border streets. Even the main streets could be not very wide, between 3 and 6 m, as the width had only to be wide enough that two packed animals could meet and pass. The widths and the length of cul-de-sacs themselves did not vary widely (1–2 m) and they were quite often curved. The street network had the appearance of being chaotic and complicated and it gives strangers and foreigners the feeling of walking within a labyrinth.	Most of the planned and newly built towns give the impression that they were standardised. They had a rectangular street layout with at least one main street. This type of town, often called Bastide, was frequently built in the southwest of France, England and Wales, Germany, Lithuania, Estonia, Latvia and Poland. But there were other forms as well with slightly curved roads and one or two main streets. Narrow streets were known as well but were not as frequent as in the Muslim cities.

continued

Table 7.1 *continued*

Muslim city	*Christian city*
Religious buildings: The Friday mosque was the centre of the town; but there could be many more mosques in the centre and there were certainly more in the neighbourhoods of the town. *Madrasas* (schools) were built from the eleventh century onwards (late thirteenth century in Morocco and Granada). They allowed teaching and learning at different levels and the highest would be university standard. Monasteries were also known in Islam under the name of *Zawiya*, *Ribats* or *Khanakan*. *Ribats* were at the beginning of the conquest also used as military camps.	The main church would mostly not have any learning facilities (but they did exist). Learning took place in monasteries or privately, and depending on the order to which the monastery belonged, learning possibilities were located either within, close to or outside the towns. Some monasteries were located along pilgrimage routes in order to house and feed pilgrims. That was also the case for Muslim towns.
Markets (*Souks***):** The streets around the Friday mosque had a wide range of shops and goods. Books and leather goods were traded in the proximity of the mosque. One product (e.g. silver goods) was normally sold in different shops in the same street. If there were many traders who produced similar products then the shops would extend to the neighbouring streets. The size of these *souks* could be very large and less valuable goods would be sold in side streets, further away from the Friday mosque. But the markets for livestock and agricultural goods were located outside the city walls (Brown, 1973: 35–36).	The early medieval town, if it was not Roman in origin, did not have a market square although space for markets was of central importance in the Christian city. The main streets were mostly used for that. From the late twelfth and early thirteenth centuries a main marketplace became popular at least in some parts of Europe. Apart from the main market, which was used for fairs, political events and presentations, but also for selling food and vegetables, there were in addition other spaces within the town for the sale of various goods. All large towns had mostly at the end of the medieval period several markets, such as for wood, hay, fruit, corn and wine. In some cities, like London, markets were mostly located outside the walls, but in most other cities they were found inside, although animal markets were often found outside as well.
Political local power: There were notary buildings and law courts for the judges (*qadis*). The *muhtasibs* supervised the markets and trade but they also controlled the general behaviour and looked after the infrastructure of a town.	The town hall and other official buildings were normally built when a town became independent of its founder(s); that happened in the late medieval period. There were offices or buildings for judges, notaries and prisons.

Muslim city	*Christian city*
Neighbourhoods and their distribution within the town: Almost all towns would have quarters of different sizes, some were so small that they consisted only of one street. Each Muslim sect would have its own neighbourhood. The same was true for ethnic strangers, people from villages and other towns, manufacturing and trading groups. If there were areas where people of different beliefs lived, they would have their own mosques, churches or synagogues. In addition, other facilities were available, such as fountains, baking ovens and baths (*Hammam*) and according to the century and ruler, non-Muslims would have their own magistrates and law courts. The rich and influential families lived in neighbourhoods close to the main mosque. The poor population had to settle next to the wall.	Neighbourhoods in the Arabic sense did not exist, apart from the areas where the Jews were allowed to reside. The rich and influential families lived close to the main square or main church or along the most important streets. The poor population and people who had dangerous jobs connected with fire or smells had to settle next to the wall.
Public spaces: Public spaces (except streets) included squares – although there were never as many as in Christian cities – and areas around fountains, mosques and green spaces. The green spaces were found inside and along the walls; existing urban orchards inside the wall would gradually be built up. Most of the large gardens were located outside. In the early centuries, cemeteries were kept inside the wall but later they were moved outside but separated according to ethnicity.	Even in the early medieval period, small gardens were known but in the late medieval period public spaces of all kinds of sizes and shapes were widespread. Within the wall sufficient areas for gardens and reserves for further buildings were provided. In the early medieval period, the cemeteries were located inside, around churches, but were later moved outside the walls, which also provided more open space within the town.
Private space: Private space consisted of small clusters of houses linked together and containing large and small open spaces. These spaces were filled up with more houses and rooms in later centuries. All houses – except the housing of the poor – had patios and this was the central and most important space for socialising with the family. The windows of the private houses faced the inner yards and not the cul-de-sac lanes; only the door gave access to the outside world. Even when opening the door, the corridor behind was built in such a way that it was impossible to see into any of the rooms. Privacy also meant that all houses and buildings, except the mosque, were built at the same height.	All houses had windows but often only small ones, and the houses of the rich had glass. Crown glass (*Putzenscheiben*) was common within the houses of the rich from the fourteenth century onwards. The poor had only shutters and possibly animal hides to protect them from the cold and rain.
Suburbs: These consisted of fruit and vegetable gardens. According to the location and the size of the suburbs all kinds of social groups resided in these outside barrios. Over time, some of these suburbs became part of the original town or they developed as neighbouring towns, like Albayzin in Granada, which became larger than Granada itself.	The Christian towns had suburbs with fruit and vegetable gardens and summer houses for the rich. As in the Muslim towns, over time some of the suburbs became part of the town and the wall would be extended. Large suburbs would have their own church and markets.

make about Max Weber's work but in this chapter, I limit myself to four other important points;

First, there are several examples, for instance Sijilmasa in the south of Morocco, which had for 200 years its own local government and the rulers emerged from a democratic process. Wirth (2000) quotes Claude Cahen (1958) who pointed out that many Islamic towns could decide for themselves and had in a sense their own local government up to the eleventh century (2000: 516). In fact, Cahen's article is an eye opener in terms of the complexity of government.

Second, Weber and many others were convinced that Islam was initially the religion of a conquering army, which was organised according to tribes and clans, and that the population had not really overcome their strong bonds to the tribes and clans and were therefore not able to identify with the city they lived in. We have already discussed this point above. This was very much in contrast to the Christian religion, which broke this connection to tribes and clans at the moment when people became citizens of a medieval town (Wirth, 2000: 22).

Third, Weber referred very little to the Arabic or to Islamic towns, and his main examples were from China, India, Mesopotamia and North Africa during the Phoenician period.

Four, the growth in local government was the result of weak imperial power in central Europe, and in the case of the Netherlands, the neglect by Philip II of economic and religious Dutch interests.

The difference in the street layouts between the two types of cities

The Muslim city

It is known that the Muslims as well as the Christians planned the majority of their towns and that the overall outlines of these towns were rectangular and the main street layouts were more or less the same. A Muslim town had, like a Christian town, three types of street – main, access streets and cul-de-sacs – and a Christian town would have main and side streets and narrow winding streets between houses – but their function was different to the cul-de-sacs. That is not to say that in both camps organic towns did not exist. Sometimes the main streets took the form of a cross, commonly known as a 'Roman' cross, although it was originally not Roman and originated in the Middle East (Figure 7.3). There is some dispute whether the rectangular form of the Muslim towns was copied from the Romans or whether it derived from much older towns in Greece, Babylon and Assyria.

But it was not only the Roman cross which was not Roman, the network of cul-de-sacs were also copied from pre-Islamic cities, as a drawing of a Babylonian residential area clearly shows (Smith, 2010: Fig. 10). At some point in history, the main and access streets became narrower – only wide enough that two donkeys or camels could pass – when the Arabs changed from wheeled to mainly

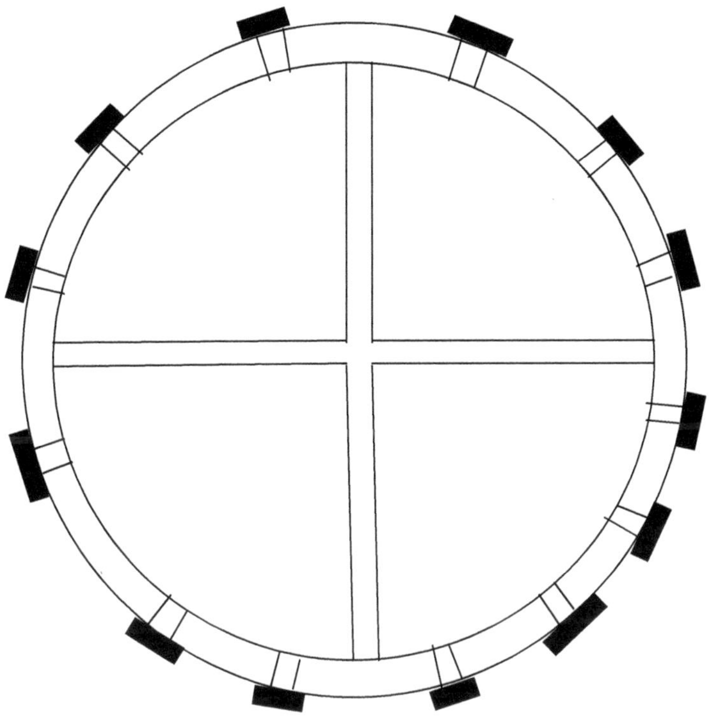

Figure 7.3 Diagrammatic representation based on the oldest found map of a city, circa 2000 BC. The city had 13 gates with a street layout in the form of a cross.

Source: Graham Crampton.

animal transport (Hourani, 2013: 124). This was connected with the time when camels had become the most important mode of transport because of their durability and the usefulness for crossing the Sahara and other deserts. This all happened before Islam had become a religion. Apparently, camels were used in the Arabian Peninsula many centuries before Christ.

An opinion about the traffic system within the Muslim cities was given by Brown (1973). He talked about the fact that the typical Arabian city was designed not to create easy access but to control and compartmentalise. He concluded that the narrow, curved streets were a built-in system of traffic control, inhibiting the movement of traffic through areas where it was best that they would not be able to go or at least be limited in numbers. In the areas of the main market, the mosque and the schools/universities, traffic access was limited. This sounds like deliberate transport planning. Is this not the same idea as pedestrianisation and car traffic restriction, which have been implemented in European cities from the 1960s onwards? Maybe the Arabs have known for thousands of years that too much traffic was bad for business?

If it is correct that most – some say all – Muslim towns were planned, how could it be that the street layout of the Muslim city has been characterised by narrow crooked streets and side streets that were nearly all cul-de-sacs? One of the reasons was that during the centuries, the street layouts were changed by their inhabitants. That was not specifically Islamic, as there is sufficient evidence that this happened in Roman cities as well in periods of less local power. In the Muslim cities it is likely that it started within the neighbourhoods because the local streets in these areas were private or semi-private and the city authority – the *muhtasib* – had officially no access to them. Hence Jewish or Christian neighbourhoods could 'design' their own space as they wanted, as could the Muslims.

Another reason, which is not really discussed in the literature, was that the population densities were extremely high and as a result every possible plot was built up, including every available street space. Although we do not have exact figures, judging by the number of people living in major Muslim towns in al-Andalus, these density figures were high compared to Christian cities at the same time. Table 7.2 shows the extremely large differences in population densities between these two types of cities. The three German cities are regarded as

Table 7.2 Estimated population size and densities in selected Muslim and Christian cities

Town	*Population*	*Size in sq km*	*Population density/sq km*	*Time*
Córdoba	Between 100,000 and 450,000 (depending on sources), with 250,000 being accepted as a likely estimate	8.00	31,250 but possibly much more (56,250)	End of tenth century
Seville	83,000	1.87	44,385	Twelfth century
Malaga	13,000–15,000	0.38	34,210–39,473	Circa tenth century
Toledo	37,000	0.77	48,052	End of tenth century
Fez	120,000	3.00	40,000	1200
London	18,000	2.59	6,950	1086
Cologne	20,000–30,000	4.01	4,988–7,481	1180
Nuremberg	23,000	1.60	14,375	Fifteenth century
Augsburg	30,000–50,000	1.73	17,341	Fifteenth century
Vienna	10,000	1.08	9,259	Circa 1200

Sources: Sources include individual city studies Povedano (2017: 83–85) for Malaga; Hirschman (2016) for German cities; and Zaimeche (2005) for Seville, plus capital city websites such as Londonist.com; Muslim historical websites such as Muslimheritage.com, plus my own calculations and observations.

the largest during the medieval period and the population figures are the first data that exist. Although the epochs are much later than the dates of the Muslim cities, one can guess that during the eleventh century the population densities in Nuremberg and Augsburg were very low. In fact, Nuremberg hardly existed at all in the eleventh century.

Another reason why Europeans find the street layout chaotic or confusing, as pointed out by Petruccioli (2002), is because they cannot recognise the original street layout any more, as it has become complex; it has quite often disappeared or half disappeared. Basically, understanding and reading Muslim cities has become difficult because these cities are different compared to the cities in the West and are often much older. Most of the authors who criticised these street layouts came from countries where wide, straight streets were the rule and experienced as 'modern'. They thought these streets were superior to Muslim streets, apart from the important detail that they are pretty uncomfortable in very hot climates. Straight, wide streets had become fashionable and were idealised during the Renaissance, largely from the sixteenth century onwards.

The Christian city

The most common form of newly planned settlements during the medieval period was the Bastide. These were built largely in two phases: the first during the eleventh century and the second between 1200 and 1400. There have been many discussions about and definitions of the word 'Bastide', what it meant and what it was. Generally speaking, these discussions are not very helpful and the words 'Bastide' and 'new town' are here used interchangeably. Bastides were new towns with their own charter and normally the inhabitants were free citizens. Many of them were simply agricultural settlements that received town status. Most of them were planned but not all and the design could vary from region to region. However, from the thirteenth to the fourteenth century, the plans of the town became more standardised, including a central square and a grid street layout (Baeriswyl, 21 August 2018 – email). Remarkably – and there is no clear answer as to why – street patterns were very similar? One can find the same design from Spain to southwest France continuing as far east as the State of the Teutonic Order (today encompassing parts of Poland, the northern part of Lithuania, Estonia and Latvia). From the mid-thirteenth century, the chequer plan with a rectangular market became the dominant street layout, not only for the larger towns but also in the small and medium-sized towns (Czaja, 2015: 242).

This brings us to the crucial question, what kind of city Mumford (1965) was talking about when he mentioned that the medieval (Christian) city was like a spider web and the curves were used to emphasise its central core (1974: 349). According to him, the typical medieval town had a street layout which took into account the contours of the landscape and had different street widths. They had mostly irregular plazas of different sizes and plenty of them. Sitte (2013) provides an answer when he mentions that the reason for the quite typical irregularity of old plazas lies in their gradual historical development (2013:

48). Maybe that was also the case with these kinds of medieval towns, which developed over centuries, and whose street networks and city shapes were both irregular. However, one cannot help but wonder whether more towns were planned than was thought possible, and often if what is regarded as typical medieval spontaneity was in reality planned.

The key characteristics of the Muslim city

The importance of neighbourhoods in the Muslim city

One of the main contrasts between Muslim and Christian cities was the ethnic mix in Muslim towns, which did not exist in the Christian municipalities, although areas where Jews were allowed to live did exist. The important commercial Christian cities had always housed different tradespeople from all over the world, such as Venice, Genoa, Lubeck, Paris, London or Bergen. But the number of these foreigners was small, often occupying only a small number of houses.

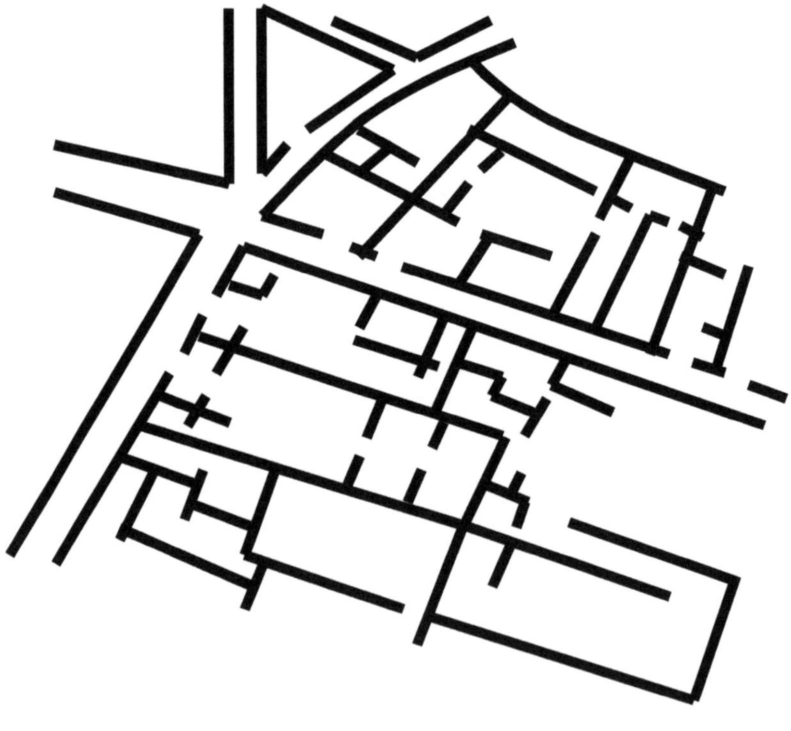

Figure 7.4 Diagrammatic representation of a neighbourhood in Ur (based on a photo by Smith (2010).

Source: Graham Crampton, based on a photo by John Smith (2010).

If one studies the neighbourhoods, which are thought to be typically Islamic, one will find that, according to a number of researchers who are summarised by Smith (2010), this type of neighbourhood was common in the major cities of Mesopotamia (Figure 7.4). Smith quoted Stone and Zimansky (2004), who studied neighbourhoods in Nippur (about 2500 BC), by writing that the differences which distinguished two neighbourhoods from each other were precisely those which defined neighbourhoods in Islamic cities (Smith, 2010: 148).

Stone and Zimansky (2004) studied another town in Mesopotamia (Maka Shapir, built about 2000 BC) and concluded that within the districts there was a mixture of wealth levels between the individual houses, and that the elite were not concentrated but broadly distributed within Mesopotamian cities (2004: 378–379). This was also true in some of the neighbourhoods of Islamic towns.

Western experts of urban sociology have often expressed the opinion that residents who live in various quarters cannot develop a real identity with the city in which they live. This could be valid for the first generation that moves to a city, but it is questionable whether this is true for the second and third generations.

What strikes me is that nobody has written about the wealth of different talents and professions that were available because they could live in separate neighbourhoods. Such arrangements made it possible for people of various backgrounds and ethnicities not only to live together but also to create a very powerful city economy on all levels, from the simple tradesmen to the more sophisticated merchants and intellectuals.

Nobody has mentioned the fact that allowing very different groups of people to live mostly peacefully within a city similarly implies that a large degree of tolerance is needed by a number of people and institutions, such as the ruler and ruling bodies as well as the citizens. Clearly, there were problems and conflicts, but by and large the structure of the city did work well most of the time. However, the aspect of safety in the neighbourhoods has been mentioned by several authors. Barrios could be closed in the night or at times of unrest with large gates. When political systems broke down, like in al-Andalus in the early eleventh century, it was not because the inhabitants of various neighbourhoods could not live together anymore; it was – as always – caused by the greed or ineffectiveness of rulers or by rivalry between them. The last substantial variation between the Christians and the Muslims was their private homes.

The construction of private space in the Muslim city

The typical Muslim house had no windows towards the streets and if so, they had to be screened. The main entrance was built in such a way that one still could not see into the house, even when the door was open, and the outside door was not supposed to face a neighbouring door directly (Othman et al., 2015: 15). Each house had a place for worship and provided privacy for women.

There was normally a separation between female and male guest areas (2015: 19). The main feature of the house was the courtyard, which included trees, flowers, plants and often a water feature. The surrounding wall of the house had to be higher than a camel rider (Othman et al., 2015). Muslim houses could have different heights. In dense urban areas they could be up to four floors in height, not including the ground floor, but all of them had to be about the same height in order to provide privacy. Some of the highest historic houses known in the Muslim world were built in Sana'a (Yemen) with at least six floors plus a ground floor. In areas of lower density the houses consisted normally of a ground floor plus one floor above. As with many other features, this type of house was likewise already known in Mesopotamia (Benevolo, 2007: 25), and a drawing of a typical house shows no window to the street (Zamarovsky, 1968: 228). The same type of house was used in the Indus civilisation of Mohenjo-Daro, 2500–1700 BC, in present-day Pakistan (Coningham, 2017: 19). Later, Greek and Roman town houses had similar features (see Hourani, 2013: 126).

Women's living quarters and veiling

The Qur'an does not say that women have to be veiled – neither their face nor their body (*burqua* or *chador*). (Compare with Mokarram Abbas' chapter on this matter.) However, the Qur'an does refer to veiling explicitly with reference to the Prophet's wives, as they were supposed not to be seen in the company of other people. The veiling of women was adopted three or four generations after Mohammed's death (632). The Muslims were copying Byzantine women of the time. Wealthier families in Byzantium had a segregated space within the house, a *Gynaikonitis*, which was a private space for women to keep men out. It seems that even this was copied by the Islamic society but it turned into something different; it became an area where men (in this case the husband) could enter (Armstrong, 2002: 14).

This brings us to a crucial question: How were men in each society and religion – whatever the religion's founders had in mind – able, in a relatively short time, to turn the situation in their favour and why did women mostly obey them? An easy answer is because the majority had no option as they were not economically independent. But in all civilisations there were powerful women, so why did they mostly go along with male dominance? There is no convincing answer. But it could have been that the inequality between men and women was not seen as a problem in societies where significant class differences were the rule; but what happened outside the family was not automatically the same as inside. This is still true today, even in Muslim culture.

Conclusion: cities and religion

Muslim towns were planned in the same way as Christian towns but the elements within the town were different. Over centuries in the Muslim towns, the street layout was changed because of increasing population densities and the existing freedom in changing neighbourhoods according to private needs. Neighbourhoods with different ethnicities were the major difference between Muslim and Christian cities. In how far local government played a major role in developing capitalism does not offer a simple answer, considering that the first industrial revolution started in England, which was not renowned for having decisive local governments. This only started from the 1830s onwards. However, there is no doubt that the bourgeoisie did play a decisive role.

The most important but also the most difficult issue is to what extent religion influenced Muslim and Christian cities? In this chapter, Muslim settlements were more the centre of analysis and the question of religion was only dealt with peripherally, as other aspects were more important in understanding these cities. As far as it is known, all the built-up elements which are today regarded as Islamic are actually pre-Islamic, though they were embraced by Islam and have not really been changed. What made a city Islamic was not the built-up environment; it was subtler and was expressed in the human behaviour and social life. Rules were provided by the Qur'an and Sharia laws. It was the way people dealt with each other, how they were able to tolerate different religions and different tribes, and how they ran their neighbourhoods in a way that was unthinkable in Christian cities.

No doubt the Christians had strict rules and codes regarding behaviour as well but I am not sure that these covered the whole society in the same way as Islam, especially after the Renaissance. There were, however, physical structures which did make clear to a visitor that one had entered a Christian or Muslim city and these were the religious buildings, mainly the mosques and the *madrasas* on the one hand and the churches and cloisters on the other.

There was also the language, the literature and the art, which were quite distinctive between the two religions. But was this true for Spanish art? And here we find a strange phenomenon, which again is not really fully understood. How was it possible that on the battlefield they killed each other, but the large houses of the rich were built in *Mudéjar* style, an Arabian design carried out by Arabian builders, in the Christian parts of Spain from the eleventh to the sixteenth century? In towns, like Toledo, the *Mudéjar* style was also used in its churches from the end of the twelfth century onwards, although it had become Christian as early as 1085. This style was used not only in churches and monasteries but also in palaces, even after these towns were conquered. Grabar (1992) has talked about the paradox of the *Mudéjar* style:

> How can one explain the contrast between policies that were leading to the destruction of Islam's presence in the Peninsula and this fascination with forms issuing from Islamic art, which continued quite consciously for several centuries, and, according to some, has remained in the background ever since?
>
> (1992: 589)

Does not the Spanish example show that our societies and especially our cities are extremely complex? Do they not all include the heritage of past and present history with its different cultures and religions? How could one think that one aspect alone, in this case 'religion', could have comprehensively formed a city even if human behaviour there is based on a specific religion?

References

Abu-Lughod, J.L. (1987) 'The Islamic city: historic myth, Islamic essence and contemporary relevance', *International Journal of Middle East Studies* 19(2): 155–176.

Alsayyad, N. (1996) 'The study of Islamic urbanism: an historiographic essay', *Built Environment* 22(2): 91–97.

Armstrong, K. (2002 [2000]) *Islam, a Short History*, London: Weidenfeld & Nicolson.

Balbás, L.T. (1970) *Ciudades Hispanomusulmanas*, 2 vols, Madrid: Ministerio de Asuntos Exteriores, Instituto Hispaño de Cultura.

Benevolo, L. (2007 [1975]) *Die Geschichte der Stadt*, Frankfurt and New York: Campus Verlag.

Brown, L.C. (1973) 'Introduction' in L.C. Brown (ed.) *From Madina to Metropolis*, Princeton, NJ: Darwin Press, 15–46.

Brunschvig, R. (1947) 'Urbanism Medieval et droit Musulman', *Revue des Études Islamiques* 15: 127–155.

Cahen, C. (1958) 'Zur Geschichte der städtischen Gesellschaft im islamischen Orient des Mittelalters', *Saeculum* 9(1): 59–76.

Coningham, R. (2017) 'Mohenjo-Daro' in J.J. Norwich (ed.) *The Great Cities in History*, London: Thames & Hudson, 18–20.

Czaja, R. (2015) 'Polish town plans as expressions of political and economic power' in H.B. Clarke and A. Simms (eds) *Lords and Towns in Medieval Europe: The European Historic Town Atlas Project*, London and New York: Routledge, 235–254.

De Planhol, X. (1968) *Les fondements géographique de l'histore de l'Islam* [The Geographical Foundations of the History of Islam], Paris: Flammarion.

Falahat, S. (2014) *A New Conceptualism of the Urban Logic of the Islamic City*, Berlin: Springer.

Grabar, O. (1992) 'Two paradoxes in the Islamic art of the Spanish Peninsula' in S.K. Jayyusi (ed.) *The Legacy of Muslim Spain*, London: Brill, 2: 583–591.

Grunebaum, G. von (1953 [1946]) *Medieval Islam: A Study in Cultural Orientation*, Chicago, IL and London: University of Chicago Press.

Grunebaum, G. von (1955) 'Die islamische Stadt', *Saeculum: Jahrbuch der Universalgeschichte* 6: 138–153.

Haleem, M.A.S.A. (2016 [2004]) *The Qur'an*, Oxford: Oxford University Press.

Hirschmann, F.G. (2016) *Die Stadt im Mittelalter*, Enzyklopädie deutscher Geschichte, vol. 84, Oldenburg: De Gruyter.

Hourani, A. (2013 [1991]) *A History of the Arab Peoples*, London: Faber and Faber.

Kennedy, H. (2014) *Muslim Spain and Portugal: A Political History of al-Andalus*, London and New York: Routledge, first published in 1996 by Pearson Education.

L'Tourneau, R. (1949) *Feś avant le protectorat: étude économique et sociale d' une ville de l'Occident muselman*, Casablanca: Publications de l'Institut des Hautes Études Marocaines XLV.

Lapidus, I.P. (1973) 'Traditional Muslim cities' in L.C. Brown (ed.) *From Madina to Metropolis*, Princeton, NJ: Darwin Press, 51–72.

Marçais, G. (1945 [1940]) 'La conception des villes dans l'Islam' [The conception of the city in Islam], *Revue d'Alger* 2: 517–533.

Marçais, W. (1928) 'L'Islamisme et la vie urbaine' [Islam and urban life] in *L'Academie des Inscriptions et Belles-Lettres: Comptes Rendus*, January–March: 86–100.

Mumford, L. (1965) *The City in History*, Harmonsworth: Penguin.

Nippel, W. (ed.) (2000) Comments in M. Weber, *Wirtschaft und Gesellschaft: Die Stadt* [Business and Community [Public and Private Life] in the City], Studienausgabe, Tübingen: J.C.B. Mohr (Paul Siebeck), 103–122.

O'Callaghan, J.F. (1975) *A History of Medieval Spain*, Ithaca, NY and London: Yale University Press.

Othman, Z., Aird, R. and Buys, L. (2015) 'Privacy, modesty, hospitality and the design of Muslim homes: a literature review', *Frontiers of Architectural Research* 4: 12–23.

Petruccioli, A. (2002) 'New methods of reading the urban fabric of the Islamicized Mediterranean', *Built Environment* 28(3): 202–216.

Pöppel, H. (2014) 'Karl der Große und Spanien: Eine zwiespältige Beziehung' [An ambiguous/uneasy relationship], Forschungszentrum Spanien [Research centre Spain], Universität Regensburg. www.uni-regensburg.de/sprache-Literatur-Kultur.

Povedano, R. (2017) *Málaga, de muselmana a christiana, la transformación de la ciudad a finales de la edad media* [Transformation of the City of Malaga: Muslim to Christian], Granada: CEA-APQ.

Sauvaget, J. (1941) *Alep, Essai sur développement d'une grande ville syrienne des origines au milieu du XIX siècle* [Essay on the Development of a Large Syrian City from Its Beginnings to the Middle of the 19th Century], Paris: P. Geuthner.

Sitte, C. (2013 [1945]) *The Art of Building Cities: City Building According to Its Artistic Fundamentals*, Eastford, CT: Martino Fine Books.

Smith, M.E. (2010) 'The archaeological study of neighbourhoods and districts in ancient cities', *Journal of Anthropological Archaeology* 29: 137–154.

Stone, E.C. and Zimansky, P. (2004) *The Anatomie of a Mesopotamian City, Makan Shapir*, Winona, Lake, IN: Eisenbrauns.

Watt, A. and Cachia, P. (1996 [1965]) *A History of Islamic Spain*, New Brunswick, NJ and London, Aldine Transaction.

Wirth, E. (2000) *Die Orientalische Stadt im islamischen Vorderasien und Nordafrika* [The Oriental City in South West Asia (the Far East) and North Africa], vol. 1, Mainz: Philipp von Zabern.

Zaimeche, S. (2005) *Seville: Muslim Heritage*, Manchester: FSTC.

Zamarovsky (1968) *Am Anfang war Sumer* [In the Beginning Was [the City of] Sumer], Leipzig : Brockhaus Verlag.

8 Women in the Muslim city

With reference to Nablus in Palestine

Mokarram Abbas with Clara Greed

Abstract

This chapter is based on a case study of Nablus undertaken by Mokarram Abbas with additional input by Clara Greed to contextualise the study and link it to key themes and religious perspectives raised in the book as a whole. Mokarram investigated the ways in which space was gendered in Nablus and how this affected women's chances of freely accessing the city. The research investigated whether Western women and planning ideas and concerns were applicable to the Muslim city of Nablus. It was found that the spatial divisions between work and home, created by Western land-use zoning, were not as evident in the ancient, close-knit, high-density, mixed-use layout of the city. Rather than having clear divisions between male and female spaces, instead she found the situation to be complex with different levels and types of division at the local level. In particular small local public parks – *mutanazahat* – *provided* women-focused oases from the harassment women encountered in the 'public' male-dominated areas. Mokarram found that women prefer their own separate spaces, and the Western idea of the complete mixing of males and females in urban public spaces and facilities was unwelcome, as this could actually reduce women's safety and freedom.

Key words: Islam, Nablus, Muslim women, land-use zoning, veils, high density, mixed uses, harassment, public parks (*mutanazahat*)

Introduction

In this chapter, Clara Greed draws strongly on the work of Mokarram Abbas, who made freely available to her the transcript of her PhD and her key journal publications (Abbas and Van de Heur, 2013; Abbas, 2013 and 2014). Mokarram investigated the ways in which women seek to inhabit, access and lead their daily lives in the modern-day Muslim city with particular reference to research on Nablus in Palestine (Abbas, 2013; Abbas and Van de Heur, 2013; Abbas 2014). Comparisons will be made with other examples of cities where religion has had a major effect in shaping and often limiting women's place in the city of man. For example Anna Papadopoulou and Juliet Davis' chapter on Greek Orthodox yields examples from Limassol in Cyprus, where some similar gender divisions of urban space are to be found. Similarities will also be drawn from other research (not the subject of a chapter in this book) such as Yamini Narayanan's work on the Hindu city, from her studies of the position of women and the role of urban planning in Jaipur in India (Narayanan, 2015: especially Chapter 7; Narayanan, 2016). The reader is advised to look at the contrasting situation described in Chapter 9 by Roja Tafaroji regarding the position of immigrant Muslim women in the non-Muslim city of Auckland, New Zealand (Tafaroji and Reeves, 2014), where Muslim women and men share urban space with all sorts of other ethnic and religious groups within a cosmopolitan, global city.

The purpose of Mokarram's research was to understand how male and female space was divided up in the city of Nablus. It is clear from her research that religion not only shapes the construction of 'sacred space', such as places of worship and gathering, but has also shaped the 'spatial' design of the urban form and the very city itself. She also sought to investigate whether the received script of Western 'women and planning' literature was relevant to the Muslim city (Hayden, 2002; 2005). Mokarram's research highlights the 'different' degrees of separation between women and men in the Muslim city. She argues that Western interpretations of Muslim urban space are over simplistic and do not appreciate the fact that rather than having an absolute public/private or male/female dichotomy in urban space, there are complex multiple layers and overlapping spaces involved, and the degree of separation depends upon the function of the space as well as the purpose of the women in that space. So she sees the city as comprising a socio-spatial continuum of semi-public and semi-private spaces. Western-educated urban planners are unlikely to appreciate this delicacy and subtlety of Muslim urban form and the advantages for women of having separate spaces from men. Mokarram considers that there is a tendency for Westerners to generalise and make assumptions about both the restrictions of Muslim cities, and what they see as 'the oppressed Muslim woman' therein (Moors, 1995; Abbas, 2014). Clearly Mokarram herself has confounded the stereotypes being a university professor, an expert in urban planning, a city councillor and a mother, which is why, being so busy, she was keen for me to organise this chapter.

Nablus

Nablus is a city of 348,000 people and one of the most important conurbations in Palestine (the administrative capital is Ramallah). Nablus is an ancient Muslim city where women's position has been clearly demarcated over the centuries within its high-density, close-knit urban form. While today Nablus' population is predominantly Muslim, it has had a very chequered history being subject to a series of conquests during its long history by a range of Pantheist, Muslim and Christian empires including the Romans, the Ottomans and the

British. Nablus today is quite a cosmopolitan city, where women have some degree of freedom, in comparison with the cities of Saudi Arabia and other stricter Muslim societies. Also, to some extent, it has been subject to aspects of modern Western city planning along with other Middle East cities (Taraki and Giacaman, 2006; Molotch and Ponzini, 2019).

As to the question of the gendered division of urban space, Mokarram found that public space is still male-dominated as a whole, especially the main squares, city centre, the coffee shops and the main thoroughfares. So women feel very uneasy about being 'out in public' if they are in the 'wrong sort' of public places. But she also found that there is some fluidity and flexibility as to which areas are viewed as male or female, especially in the local neighbourhood areas away from the centre. This situation is similar to that described by Anna Papadopoulou 3under Greek Orthodox Christianity, where there are unwritten codes as to how, where and why women use public space. In the light of such studies, much accepted urban design theory in which only one 'image of the city', and a male one at that, is presented as being relevant to 'everyone' needs to be re-evaluated (Lynch, 1968; Madanipour, 1996; Carmona et al., 2003). Arguably there is no one view or experience of the city but different ones, according to the gender and religion of the observer, the locality and the level of religious tradition and observance in the city in question (Abu-Lughod, 1987)

The division of space between men and women is not as clear cut or 'absolute' as some Western commentators would suggest when discussing 'the Muslim city'. As will be explained, Mokarram found that there are degrees of types of gendered space, from the exclusively female indoor and courtyard areas, to the shared public spaces in the urban parks and the more indeterminate, still evolving areas where education, shopping, business, commerce and recreation take place. Whether women are 'allowed' in a particular urban space depends to a considerable degree upon why they are there, for example whether they are visiting relatives and friends, shopping or going to weddings, graduations, religious events or funerals. But in both Mokarram's and Roja Tafaroji's studies there is little mention of women going out on their own; they are always with others, with their family or in groups. There seems to be little concept of a woman having the freedom, or the personal inclination, to wander around alone, like the classic male *flâneur* who finds the city a place of emancipation (Amin and Thrift, 2004) (and compare Gill Reeves' observations (Chapter 13) on women's often precarious 'place' in the night-time city)

Mokarram's study concentrated upon looking at two main case studies of small urban parks – as well as looking at the city as a whole – which are called Mutanazahat (متنزهات Mutanazah is the Arabic word for such a park), which can provide a welcome oasis for women in the city of man. These are not parks in the English sense of large areas of grass and flowers. Rather they are small, structured areas, often with seating, trees and hard-paved areas, that also provide small recreation areas, entertainment and often refreshment kiosks and children's play areas. The two she studied were first the Jamal Abel Nasser Park, which was popular with women, but was not exclusively for women, so they had to keep to their allocated part of the park and were normally accompanied by children. But it worked both ways in that individual men straying into the women's area were generally looked upon with suspicion and shouted at to move out. The second park was the Children's Happiness Centre (Abbas and Van de Heur, 2013: 10), which was specifically designed for women (Nablus Municipality, 2008). These parks give women 'the right' and 'excuse' to be in public space. But women still encountered harassment, described by the Arabic term Zor'an (زعران). Zor'an broadly means men behaving inappropriately towards women in public space and breaking the codes of proper social behaviour, what we would call harassment (Abbas and Van De Heur, 2013: 15). This might involve shouting at women, entering women's spaces, simply being rude and intrusive and basically staring at women. There is expected proper behaviour for both men and women, as evidenced in the city parks. Men may be condemned for coming into women's areas even if they are accompanied by children. But women may be shouted at by the men if they are seen as being too noisy or just playing on the swings. To avoid problems women would feel uneasy about sitting on the same public seats as men or walking along pavements in male-dominated locations. Interestingly a lot of effort has gone into protecting women in the public urban space (rather than addressing male behaviour), for example, cash machines often have separate queues and facilities for women and men.

What is the problem? The Western feminist argument

One of the purposes of Mokarram's research was to see if Western urban feminist literature on the 'problem of the city of man' was applicable to Muslim cities. Therefore I need to digress briefly to explain 'the problem'. It has long been argued that the way in which cities have traditionally been designed by a predominantly male town planning profession, in North America, the UK and Europe (McDowell, 1983; Greed, 1994; Hayden, 2002; Greed and Johnson, 2014: Chapter 15) has not taken into account the needs of women. In particular the emphasis upon trying to create a rational, orderly pattern of land use through land-use zoning has been challenged. Zoning has separated residential areas from employment areas, and made it difficult for women to work outside the home while also still undertaking childcare and all the myriads of home-making duties for which they are responsible. Likewise the location of employment zones has not been conveniently designed in relation to shops, schools and other necessary amenities (Madanipour, 2004). Therefore the rationality of zoning has not taken account of women's 'different' way of life, and, arguably, far from being modern and scientific, in my own previous research I found that the very concept of zoning was deeply laden with ancient superstitious attitudes regarding controlling women's place in the city of man (Greed, 1994: 79–82) with an emphasis upon women being part of the private realm 'inside' the home, and men being part of the outside public realm of the city.

All these problems have been reinforced in the West by transportation systems which, until relatively recently, have favoured the motor car, to which fewer women have access (Greed, 2019). While town planners always saw the importance of public transport too, much of the provision has been concentrated around meeting the needs of the male commuter during the rush hour, with the emphasis upon straightforward journeys from the suburbs to the city centre or to the district centres and other employment areas (Greed and Johnson, 2014: Chapter 7). The women and planning movement has highlighted the problems women have trying to get around cities designed primarily to accommodate male travel patterns. Much of women's employment is part-time, takes place in district centres rather than in the city centre, and their journeys take place outside the rush hour. Also women's journeys are much more complicated as they seek to combine their journeys with stop-offs along the way: as women trip chain their journeys. For example, a woman may set off from home dropping her small child off at the childminder (day care), then take her older child to school and then travel to work. On returning she may repeat this journey in the other direction, plus probably having to go to the shops, for food, and maybe visit the doctor's surgery, the post office or other necessary destinations. Many of such journeys are short and they are not direct radial trips to and from the centre; rather they are likely to be tangential, often between neighbourhoods and local districts – journeys that are often un-provided for by public transport systems.

Solutions to these problems put forward by women planners include proposing new types of 'city form' that will recognise and accommodate women's 'different' needs better, and are not limited to local street level issues such as building more ramps, improving street lighting and creating safer city streets (although these issues are very important too.) For example, European women planners have campaigned to create 'the city of short distances, the city of many local centres, and the city of walking and adequate public transport' (Sanchez de Madariaga and Roberts, 2013). While some of these goals tie in with male ideas regarding creating sustainable cities, the detailed specifications are likely to be somewhat different, particularly in relation to accommodating women's different travel patterns and their need to combine both work and home duties (Fainstein and Servon, 2005).

Mokarram found that the Western feminist analysis of the city was not necessarily relevant. Middle Eastern cities have a much more close-knit urban form, but many more mixed neighbourhoods with all sorts of people living close together, albeit with very subtle social class and status demarcations between neighbourhoods, which like the complex layers of difference, and nuances of meaning, between male and female spaces, can only be fully appreciated by residents fully cognisant of the culture (Wright, 1981; Mustafa, 2010). It is only in areas where modernisation and new residential developments on the outskirts of the city have taken place that the problems of separation of work and home and zoning come into play (Al-Lail, 1996).

Even in the West, a level of caution must be introduced, as different classes and types of women are affected in different ways by zoning policy, and may not experience the same problems as described in the main 'women and planning' literature, but they may not have a 'voice' to say so (Darke et al., 2000). Some working-class women and men, and increasingly poor and immigrant urban populations in Western countries, especially those living in poorer areas which in the UK at least were often located around the city centre, often find work close to where they live in the city centre, especially within the service sector. So they may not experience a major commute from the suburbs to their workplace, unless they have been 'moved out' to distant social housing estates as part of urban clearance programmes. Furthermore, in many continental European cities, such as Paris, it is not the poor who live nearest to the city centre, but the rich in very expensive apartments in prime *arrondissements* (expensive city centre districts), whereas it is the poor, the immigrant and the remnants of the traditional old working class that are moved out into the *banlieues* (suburban social housing estates) on the edges of the city (Dolors et al., 2004; Madanipur, 2004).

In fact the planners themselves have lost faith in zoning to some extent and the extended home to work journeys that it creates. As discussed in my continuing work on 'women and transport' (Greed, 2019) nowadays planners are more obsessed with environmental sustainability and have become disenchanted with the motor car. But there has been little consideration of the impact of transport policies that favour cycling upon women, particularly those who have to travel long distances accompanied by small children. (Few women cycle in the Muslim world and many in the West have trepidations because of both road safety and personal safety reasons.) There is also a whole range of issues highlighted over the years by Western women urban planners regarding issues of personal safety and street harassment, poor public transport, the lack of accessibility of many city streets for those with push chairs or wheelchairs, and also the lack of absolutely basic facilities such as public toilets (Ramster et al., 2018).

The different situation in Muslim cities

While many Muslim cities have not, yet, been affected by the zoning problems discussed in Western 'women and planning' literature, zoning divisions of cities were not seen as necessarily 'bad' by Mokarram and the women she interviewed. Because of the predominance of men in the main public spaces, she found women actually welcome having separate female spaces as a strategy to counteract male space domination and as a means for the city to function for them too (Abbas and Van de Heur, 2013: 156). Rather than seeing segregated spaces as a sign of patriarchal domination, in reality they were often fought for and gained through action and campaigning by women themselves, to carve out their own spaces in the city of man. Women-only spaces were seen as places that can empower women, albeit a pragmatic strategy, in male-dominated urban societies (Crawford et al., 2011).

An emphasis upon the Western problems of city-wide approaches to zoning did not really help address the issues around the local level of use of space in complex, finely-grained, close-knit neighbourhood locations, where much of many Muslim women's lives was centred. As stated earlier, Mokarram found that many women still predominantly see the indoors as their main realm, and that men dominate the outside public realm. But she also saw a series of multiple layers to this situation, with a range of semi-public and semi-private areas particularly within the local, more 'hidden' residential areas and neighbourhoods, where women had greater control over public space. There are many secluded courtyards and footpaths that are hidden from 'public view' and that are frequented by women within and alongside the curtilage of their home. Also many Middle Eastern houses are surrounded by blank high walls, but once inside they often contain large courtyards which provide high levels of privacy for women and children. Unlike many Western cities with their clear, direct wide roads through the city and a high level of legibility in finding one's bearings, as Mokarram explains (and Carmen Hass-Klau has already explained) many Muslim cities include 'msattar: مستّر', that is enclosed spaces without seemingly having any way through, that is without any clear visual corridor attaching them to the rest of the city. Such designs have often given an impression of 'impenetrability' to would be invaders and lost tourists (Mortada, 2003). The only way to navigate is to look up and look for landmark mosque minarets.

In her study of women's space in Cyprus, as set forth in Chapter 5, Anna Papadopoulou also found that there were differing degrees of 'publicness' and 'privateness' in Limassol, with the curtilage around the front of the house and also the balconies on the front façade providing protected areas for women. Some such balconies were closed in, but still had a view out, an architectural legacy from the Turkish Muslim period of Cyprus' history. She also found that rather than there being clear-cut male/female delineations in the neighbourhood areas, women had their own 'different' footpaths and routes which meandered through the housing areas and avoided the main thoroughfares and squares which have often been commandeered by men. In both Limassol and Nablus men tended to sit at the street-side (male-only) cafés on the main squares and boulevards watching the passers-by ready to call out at women who were 'out of place'. It is interesting that unlike in the West, where places where alcohol is consumed (pubs, bars) have tended to be male-dominated, in the nominally non-alcoholic Muslim countries cafés perform the same function.

Narayanan (Narayanan, 2015; 2016), in studying historical Hindu cities in India, in particular in Jaipur, found similar complex, multi-use, gender-mixed neighbourhoods. New gendered problems develop when modernisation and especially Westernisation takes place opening up the city and making women more vulnerable because of loss of the backstreets, footpaths and small squares that they used to access (and cut through) the city of man. Many Indian cities have been undergoing a great deal of demolition to make way for new Western-style development. Narayanan found that so-called slum clearance has often completely destroyed complex community structures and urban formats that have grown up over centuries. Such neighbourhoods were often focused around local temples, many of which are being razed to the ground. But in tourist areas she also noted that some neighbourhood temples and the ancient housing around them were being renovated and their quaint architecture conserved to attract Western tourists. In this process the local people are likely to be moved out and their temples become 'museums' for tourists to photograph. Other historical buildings may even be turned into expensive hotels for Westerners who want to experience 'the authentic Indian experience'. This process has not happened yet in Middle Eastern Muslim cities, mainly because of their unpopularity with tourists because of perceived terrorist threats, and uncertainty among Western women tourists as to whether they need to cover up.

Not only were the buildings and secluded spaces for women demolished but the traditional gendered pedestrian route-ways between the buildings were also obliterated. Narayanan noted that women were unlikely to use the new Western-designed roadways and boulevards, because they did not go where they needed to go, footpaths were often minimal relative to the space given to the main carriageway for cars, and also they would feel very vulnerable and exposed to the male gaze on such new roads, from both drivers and other pedestrians. Women felt safer avoiding such visual exposure.

But in terms of the purpose of women's journeys, and the chances this gave them to penetrate public urban space, both Narayanan and Papadopoulou found the importance of going to the local temple or church, respectively, a very important reason for 'getting out of the house', using secluded footpaths and backstreets to do so, not male-dominated open roads or modern boulevards and squares. It would seem that men knew they were not meant to harass women who are making religious journeys to the small back-street temples, (who in Jaipur can be seen to be carrying garlands and other offerings for the gods). Narayanan notes that such small local temples became safe meeting places for local women as well as sites of worship. Papadopoulou found likewise that Greek Cypriot women frequented the smaller back-street chapels and churches, away from the male gaze. Maybe this is also similar to wearing a veil in Nablus.

Veils and accessing urban space

This is an apt point in the discussion to raise the issue of the veil (Hoodfar, 2001; Bucar, 2017). Mokarram explains that there are different interpretations of why some Muslim women wear veils and its function. Some Westerners see it as a sign of oppression, whereas some Muslim women see it is as a sign of freedom from Western influences and an over-sexualised culture. Whatever a woman's real opinion, Mokarram points out that wearing a veil can be a very useful way of enabling women to penetrate male areas of the city, by rendering themselves invisible. She argues that the separation of women and men, in different buildings, urban spaces and so forth, can actually benefit women, given the intrusive and often threatening way that many men look at women. The separation of male and female can be extended to other goods and services, including the education system (Leathwood, 2004). But Mokarram makes it clear that she herself believes that the Qur'an endorses the equality of women (Qur'an Chapter 49:13: Surat Al-Hujurat (The Rooms)) and she addresses the 'difficult passages' which men have used to confine women, when in fact the Qur'an says that men should respect women too and not stare at them but 'lower their gaze' (as explained in Annexes 3–5 of her PhD). I have been through all this myself regarding what the Bible says about women, investigating the validity of those verses that say women should keep silent, and weighing them against those verses which demonstrate that women are speaking, teaching and being very active in all aspects of the early Christian community (see Chapter 6). My own personal maxim when seeking to interpret the ostensibly more sexist aspects of scripture is to 'Read all the verses', and not to pluck individual verses out of their wider textual setting.

Mokarram objects to Western interpretations, assumptions and conclusions as to the use of the veil, and also its conflation with the idea of purdah, which may range from women staying behind a veil or screen when in the company of males, to women being secluded indoors most of the time (Papanek, 1973). Purdah does not have to be either religious or Muslim in origin and varies from country to country. For example, Narayanan explains that high caste Hindu women also used to wear a veil and some families subscribe to purdah. Women staying at home, in purdah, and not having to work, is seen as a sign of being high caste (Mandelbaum, 1998). Some Hindu women will wear a head scarf called a *ghonghat* when they leave the *haveli* (the confines of the traditional house) to go to work or to the temple. Interestingly lower caste women such as field and factory workers are less likely to wear a veil or head scarf, except when they visit a temple.

In comparison, the problem of harassment (*Zor'an*) and gender segregation is widespread in Western cities too, although it might be somewhat hidden or simply dismissed as culturally 'normal'. Indeed the old-fashioned, but far from extinct idea of a woman being 'just a housewife' who does not have to work because she is supported by her husband is almost a form of purdah. Western cities are not as visibly divided into male and female areas, although arguably divisions do exist between public and male/female private spaces. Nevertheless many women 'self-police' when walking in the city on their own, especially in remote backstreets, areas around male-dominated pubs and bars, and many parts of the city in the evening and night-time. Indeed many localities and buildings are still not welcoming for women, and women are likely to feel out of place or to be challenged for even being there, for example in the case of some sporting venues, certain pubs, red light districts and sex clubs. In fact until relatively recently most pubs, particularly in working-class areas, were seen as male territory and it was very unusual for a woman to enter one on her own. Many only provided male toilets and under-age children were not admitted thus effectively barring those women who wanted to have a family night out with their husband. But any woman can still feel unsafe anywhere in the city of man, in any location, especially in public transport systems, at desolate railway stations and over-crowded trains filled predominantly with male commuters. Ironically those women that stand out as different in Western cities, for example Muslim women wearing hijabs to discourage harassment, may actually attract rather than deflect unwanted attention, because they stand out as 'different'. But, as will be explained later in the chapter by Roja Tafaroji, her research in New Zealand found that Auckland was such a diverse city, with immigrants from many races and religions, plus of course Maori citizens, and all sorts of other groups that expressed their sexuality or life style through their dress, that nobody gave a second glance to Afghan women as they went about their business in the city.

Conclusion

It is important not to apply Western planning principles to Middle East cities as their morphology and principles are quite different, and a variety of religious and pre-secular cultural factors need to be taken into account, especially when 'planning for women' (Moser, 1989; 1993). Likewise one must be careful not to 'condemn' Muslim women for their particular way of life, as they are adopting realistic solutions which will enable them to access the city of man and to carve out for themselves their own space. This is particularly so in their control over courtyards and secluded paths within their neighbourhoods at the local level (UNFPA, 2011). Nor should it be assumed that mixed spaces and facilities are necessarily always better for women, either in the East or in the West (Ramster et al., 2018), Arguably in the Muslim city women have acquired more control over certain parts of the built environment than is the case in some Western cities. By comparison, in the UK, in theory, anyone can go anywhere but women are always somewhat wary and careful when entering certain public spaces, for as Doreen Massey the urban geographer commented, 'to be male is to occupy space' (Massey, 1994).

References

Abbas, M. (2013) 'Women's strategic needs in outdoor recreational spaces: the case of Nablus, Palestine', Paper presented at Athens Institute for Education and Research Conference Series MDT2013-0442. www.atiner.gr/papers.htm.

Abbas, M. (2014) *Thinking Public Spaces through the Spatiality of Arab Women: The Case of Nablus, Palestine*. PhD thesis published by VUBPRESS, Brussels University Press, Vrije Universiteit Brussels.

Abbas, M. and Van de Heur, Bas (2013) 'Thinking Arab women's spatiality: the case of "mutanazahat" in Nablus, Palestine', *Gender Place and Culture* 21(10) (2014): 1214–1229, published online: 18 July 2013 at, http://dx.doi.org/10.1080/0966369X.2013.817969.

Abu-Lughod, Lila (1987) 'The Islamic city: historic myth, Islamic essence, and contemporary relevance', *International Journal of Middle East Studies* 19(2): 155–176.

Al-Lail, H.R.J. (1996) 'Muslim women between tradition and modernity: the Islamic perspective', *Journal of Muslim Minority Affairs* 16(1): 99–111.

Amin, A. and Thrift, N. (2004) 'The "emancipatory" city?' in L. Lees (ed.), *The Emancipatory City? Paradoxes and Possibilities*, London: Sage, 231–235.

Bucar, E. (2017) *Pious Fashion: How Muslim Women Dress*, London: Harvard University Press.

Carmona, M., Heath, T., Taner, O. and Tiesdell, S. (2003) *Public Places – Urban Spaces*, London: Routledge.

Crawford, Catherine, Miles, Debra and Plumb, Diane (2011) 'Women-only space facilitating empowerment' in Peter Jones, Debra Miles, Abraham Francis and S.P. Rajeev (eds) *Eco-social Justice: Issues, Challenges and Ways Forward. Voices from De Novo '11*, Karnataka: Books for Change, 110–125.

Darke, J., Ledwith, S. and Woods, R. (2000) *Women and the City: Visibility and Voice in Urban Space*, Basingstoke: Palgrave.

Dolors Garcia-Ramon, Maria, Ortiz, Anna and Prats, Maria (2004) 'Urban planning, gender and the use of public space in a peripheral neighborhood of Barcelona', *Cities* 21(3): 215–223.

Fainstein, Susan and Servon, Lisa (2005) *Gender and Planning*, London: Rutgers University Press.

Greed, Clara (1994) *Women and Planning: Creating Gendered Realities*, London: Routledge.

Greed, Clara (2019) 'Are we still not there yet? Moving further along the gender highway' in C. Lindkvist Scholten and T. Joelsson (eds) *Integrating Gender into Transport Planning*, Cham: Springer Nature, 25–42.

Greed, Clara with Johnson, D. (2014) *Planning in the UK: An Introduction*, London: Palgrave Macmillan.

Hayden, D. (2002 [1982]) *Redesigning the American Dream*, New York: Norton (2nd edition).

Hayden, Dolores (2005) 'What would a nonsexist city be like?' in Susan Fainstein and Lisa Servon (eds) *Gender and Planning*, London: Rutgers University Press, 47–64.

Hoodfar, Homa (2001) 'The veil in their minds and on our heads: veiling practices and Muslim women' in Elizabeth Castelli and Rosamond Rodman (eds) *Women, Gender, Religion: A Reader*, New York: Palgrave Macmillan, 420–447.

Leathwood, Carole (2004) 'Doing difference in different times: theory, politics, and women-only spaces in education', *Women's Studies International Forum* 27(5–6): 447–458.

Lynch, K. (1968) *The Image of the City*, London: MIT Press.

Madanipour, Ali (1996) *Design of Urban Space: An Inquiry into a Socio-spatial Process*, New York: John Wiley & Sons.

Madanipour, Ali (2004) 'Marginal public spaces in European cities', *Journal of Urban Design* 9(3): 267–286.

Mandelbaum, David G. (1988) *Women's Seclusion and Men's Honor: Sex Roles in North India, Bangladesh, and Pakistan*, Tucson, AZ: University of Arizona Press.

Massey, Doreen (1994) *Space, Place and Gender*, Cambridge: Polity Press.

McDowell, L. (1983) 'Towards an understanding of the gender division of urban space', *Environment and Planning D: Society and Space* 1(1): 59–72.

Molotch, H. and Ponzini, D. (2019) *The New Arab Urban*, New York: New York University Press.

Moors, Annelies (1995) *Women, Property and Islam: Palestinian Experiences, 1920–1990*, Cambridge: Cambridge University Press.

Mortada, Hisham (2003) *Traditional Islamic Principles of Built Environment*, London: Routledge.

Moser, Caroline (1989) 'Gender planning in the third world: meeting practical and strategic gender needs', *World Development* 17(11): 1799–1825.

Moser, Caroline (1993) *Gender Planning and Development: Theory, Practice and Training*, London: Routledge.

Mustafa, Osama (2010) *Morphology of Spaces and Public Squares in Old City of Nablus: Analysis and Suggestions of Development*. Unpublished Master's dissertation (in Arabic). Nablus: An Najah National University, Department of Architecture.

Nablus Municipality (2008) *Children Happiness Center: Introduction*. www.nablus.org/en/Children-Happiness-Center-01.php (accessed 8 March 2012).

Narayanan, Y. (2015) 'Women's right to the city: integrating Hindu religion and heritage' in *Religion, Heritage and the Sustainable City: Hinduism and Urbanisation in Jaipur*, London: Routledge, 159–183.

Narayanan, Y. (ed.) (2016) *Religion and Urbanism: Reconceptualising Sustainable Cities for South Asia*, London: Routledge.

Papanek, Hanna (1973) 'Purdah: separate worlds and symbolic shelter', *Comparative Studies in Society and History* 15(3): 289–325.

Ramster, Gail, Greed, Clara, Bichard, Jo-Anne (2018) 'How inclusion can exclude: the case of public toilet provision for women', *Built Environment* 44(1): 52–77.

Sanchez de Madariaga, I. and Roberts, M. (eds) (2013) *Fair Shared Cities*, London: Ashgate.

Tafaroji, R. and Reeves, D. (2014). 'The everyday life of immigrant women as a struggle to create their own living places', Paper presented at *Designing Sustainable and Inclusive Urban Environments*, Gender STE Engendering Cities, Rome, 24–25 September.

Taraki, Lisa and Giacaman, Rita (2006) 'Modernity aborted and reborn: ways of being urban in Palestine' in Lisa Taraki (ed.) *Living Palestine: Family Survival, Resistance, and Mobility under Occupation*, Syracuse, NY: Syracuse University Press, 1–50.

UNFPA (2011) 'Nablus Municipality: a successful model in women's empowerment' *This Week in Palestine*, July (159): 28–29. Feature on a Report from the United Nations Population Fund (UNFPA). www.thisweekinpalestine.com/i159/pdfs/July%20159-2011.pdf.

Wright, Susan (1981) 'Place and face: of women in Doshman Ziara, Iran' in S.G. Ardener (ed.) *Women and Space: Ground Rules and Social Maps*, London: Taylor & Francis, 136–156.

9 Afghani Muslim women in Auckland, New Zealand

Making places

Roja Tafaroji

This chapter is dedicated to the memory of 50 precious lives, who were taken from us on the 15 March 2019 while practising their religion and worshipping in their mosques in Christchurch, New Zealand. May this chapter be among the thousands of documents and discussions that contribute to the fight against ignorance.

Abstract

Urban policy making and academic research tends to categorise the needs of different groups of people in society, based on one aspect of their identity, such as gender or ethnicity, but often ignores the effects of the intersection of other key aspects of people's lives and identities, such as their religion, race and stage of life among others (Crenshaw, 2019). Based upon Roja's fieldwork, this chapter challenges the mainstream discourse by emphasising the importance of religion in a collective culture like Afghan as a key factor in generating different urban life experiences for women (as against men) in a Muslim community. The chapter incorporates material from detailed interviews regarding the personal and spatial identities of eight Muslim Afghan women from a refugee background living in the global city of Auckland, New Zealand. It discusses the intersection of their affiliation to the Muslim religion, their Afghani origins, within the Hazara ethnic community, as well as their gender, and how these shape their daily lives. This discussion is set against the wider 'colonial discourse' and 'Western' attitudes towards migrants, which also regulate and shape their experiences. Through a range of semi-structured interviews, cognitive mapping and observation of daily activities and community group participation of eight Afghan women in Auckland, this research reveals a range of levels, layers and types of discourse of Afghan women about their right to the city.

Key words: community, religion, ethnicity, gender, Muslim Afghan women, intersectionality, Auckland New Zealand

Introduction: the intersection of religion, gender and ethnicity

While the last two chapters discussed urban form in Muslim cities, with Mokarram Abbas looking at the position of women therein, in contrast this chapter looks at the position of Muslim women in a non-Muslim Western city in relation to their collective culture. The research is focused upon Muslim Afghan women who have migrated to New Zealand, where the norms and values are different from the ones of traditionally Muslim societies (Tafaroji, 2014; Tafaroji and Reeves, 2014). The study is based upon the global diverse city of Auckland, where immigrant Muslim women potentially must have more spatial opportunities to access and enjoy the city. However, as will be explained further, Afghan women in this study express their representations in the city influenced by different layers of their identities, by different sources of power within geographies of encounter. While Carmen Hass-Klau's chapter took a city-wide view, and Mokarram's research included a considerable number of Muslim women in Nablus, this chapter is focused mainly – within the research approach and methodology – upon the detailed examination of everyday life experiences of eight immigrant Muslim women from Afghanistan.

The research significance in Auckland

In order to achieve equitable and just urban policy making, which recognises the needs of a very diverse range of social groups in a global city (Spoonley and Butcher, 2009), including the Muslim Afghan women studied, it is important to understand the differences between people, their defining characteristics and their individual and collective values which shape their sense of identity (Fraser, 1998; Young, 1986; 1990; 2000; Nicholson, 2008). Gender, religion, ethnicity are among the most important aspects of identity, and these factors need to be recognised to achieve an equitable approach to urban planning policy within diverse global cities. Urban planning research and policy, however, has generally failed to acknowledge these differences particularly in relation to the needs of immigrant women from different religious and ethnic backgrounds in their new host cities. While the majority of scholarship has focused on simplified male/female gender differences, there is an apparent ignorance about the modulating effects of religion and ethnicity on specific women's and men's lives, and how these factors intersect spatially in the current era of globalisation (Beebeejaun, 2017; Fenster, 1999; Jarvis et al., 2009; Kail and Irschek, 2013; Mohammad, 2005).

The migration pattern of Afghan refugees as well as the resettlement process in New Zealand, as with other immigrants and refugees with a Muslim background, has been influenced by the negative nature of reports in the Western media regarding Muslims. The emphasis has been upon the threat of terrorism, and issues of national security, particularly since 9/11 (11 September 2001) (Ahmad, 2006; Ahmed and Matthes, 2017; Ewart et al., 2016; Lahav, 2003; Kabir and Bourk, 2012). Particularly in the cases of Afghan refugees, the mainstream media in Western societies presents an image of Afghanistan and their people, specifically Afghan women, as oppressed 'others' who need to be saved by those living in Western societies (Abu-Lughod, 2002). Muslims have been portrayed as 'Orientals' characterised as untrustworthy and dangerous: as the 'Other' within Western society (Said, 1993; 1997). Thus the West (re)produces a colonial discourse to define itself as the controlling power over the East, of First World and Third World (Mohanty, 1988; Ahmed and Matthes, 2017; Ewart et al., 2016; Said, 1997; Spoonley and Butcher, 2009).

Afghan women in Auckland

According to Census 2013 in New Zealand, the number of people who identified themselves as Afghani was 3,414. Around 25 per cent of this population were born in New Zealand. Ninety per cent of the overseas-born population were born in Afghanistan (Statistics New Zealand, 2013). In 2013, around 71 per cent of the Afghani population lived in the Auckland region (2,424 people) (Statistics New Zealand, 2013). Afghan women living in Auckland make up almost half of this population.

Research about the situation of Afghan women as immigrants with a refugee background in Western countries suggests that although they may change their socio-spatial settings, they still tend to carry on their traditional roles and responsibilities in their families and communities (Bhanji, 2011; Khan, 2002; Lipson and Miller, 1994; Rostami-Povey, 2007; Malkki, 2019; Barfield, 2010). Historically, this is a result of their life experiences under the influence of the patriarchal culture in Afghan families. These experiences have been aggravated by their experiences of exile as a result of civil war in Afghanistan from 1992 to 1996, and when the Taliban captured the capital city of Kabul in 1996 (Baker, 2010; Bhanji, 2011; Ghasemi, 1998). Following a century of war, exile and (economic, social and ideological) oppression, many of these women left their hometowns and country to seek a 'better' life (Clements, 2003; Marsden, 2001; Dupree, 1980). On the other hand, they experience their daily lives in the context of Western countries where globalisation is evident and (neo)liberal policies promote individualism and personal identity as well as independence in people's daily lives (Elliott and Lemert, 2009; Woodward et al., 2008).

The Taliban (Pashto: طالبان *ṭālibān*) is an Islamic fundamentalist political movement in Afghanistan. It spread from Pakistan into Afghanistan and formed a government, ruling as the Islamic Emirate of Afghanistan from September 1996 until December 2001, with Kandahar as the capital. While in power, it enforced its strict interpretation of Sharia law, and leading Muslims have been highly critical of the Taliban's interpretations of Islamic law. Women were affected by the Taliban's interpretation of Islam in a cruel way. They became confined to their homes to engage in domestic work. Under the Taliban's harsh version of Islamic law, women were not permitted to work outside their homes. Even their presence in public spaces, such as when shopping and visiting relatives' places, resulted in many limitations and obstacles if they were not accompanied by a male (Baker, 2010; Bhanji, 2011; Ghasemi, 1998; Khan, 2002).

Because of this background, gender, ethnicity and religion are regarded as imperative in the identification process of Afghan women (Rostami-Povey,

2007; Gupta and Ferguson, 1992; 1997; Crenshaw, 2019). They contribute to different experiences of Afghan women in their new habitats.

Gender

A technical report published by Auckland Council (Gilbertson and Maeres, 2013) suggests that gender identity raises controversial issues for both men and women in Auckland with a refugee background. 'Women had difficulties being a sole parent and with differences between their culture and New Zealand culture, such as in dress and the role of women. Men had problems with health issues, finding a partner, loneliness, racism and discrimination' (Gilbertson and Maeres, 2013: 39). The division of labour between men and women also raises challenges that different gender groups experience in urban society. As mentioned earlier, Afghan women carry traditional roles and expectations according to their cultural norms. Suburbanisation and zoning, associated with the traditional urban growth pattern in Auckland (Austin et al., 2014; Dixon and Dupuis, 2003; Memon, 1991; Saville-Smith, 2000), have divided housing areas from working areas, which supports the division of labour between men and women. Traditionally, housing and working areas are respectively known as places of women and men (McDowell, 1997). The suburbanisation and zoning policies and regulations, thus, have strengthened the traditional gender roles within the Afghan population living in Auckland. However, recent urban developments in Auckland support intensification and mixed-use developments in suburbs, which may provide working opportunities outside their houses for women from cultures like Afghanistan. Furthermore, gender roles are expected to be different for different generations due to changes in life-cycle. This will be described further.

Ethnicity

Statistics New Zealand provides a definition of ethnicity as 'a measure of cultural affiliation, as opposed to race, ancestry, nationality or citizenship' (Statistics New Zealand, 2013). A sense of belonging to one or more than one ethnicity contributes to people's identification process. This sense of belonging provides people of the same ethnicity with emotional and effective connections through their shared memories and culture (Marlowe, 2015). Depending on the common characteristics between people of one ethnicity, there may be many differences and similarities between the ethnicity of migrants and those of the majority in the host society. Growing up in a collective (tribal) culture (Khan, 2002; Rostami-Povey, 2007) and living in an individualistic (liberal) society (Beaglehole, 2013; Liev, 2008; Ross-Sheriff, 2006; Thorns et al., 2010; Ward and Masgoret, 2008), Afghan women who live in Auckland face the cultural expectations of the Afghan community versus those of the host society. These different expectations sometimes confront Afghan women with conflicting situations. Conforming to either of the cultural expectations may lead them to assimilation in and isolation from one of the cultural structures.

Religion

Statistics New Zealand suggests that religion 'guides people's conduct and morals' (Statistics New Zealand, 2014: 31). Everyday life experiences of Afghan women, then, are directed through their religious values.

Muslim communities, like other religious communities, play an important role during the resettlement process of (new) immigrants in the host society. The migration pattern of Afghan refugees as well as their resettlement process, like for other immigrants and refugees with a Muslim background, has been highly influenced in New Zealand, as in other Western countries, by political events as well as the mainstream Western media, particularly since 9/11 (Ahmad, 2006; Ewart et al., 2016; Lahav, 2003). The Western media has been highly influential in presenting an image of Muslims that maintains an Oriental view of threat and untrustworthiness: defining the 'Other' in order to define itself (Said, 1993; 1997).

Key characteristics of Afghan Muslim women in this research

Table 9.1 shows the main characteristics of the participants' identities to provide a foundation with which to understand their stage of life at the time of the study. The study was undertaken in two phases, across five years, with a short gap in the middle, which is why there are references to the first and second phases of the study in Table 9.1.

The table shows that there are some similarities as well as differences between the participants of the research. It demonstrates some of the quantitative aspects of the participants' personal and collective identity which is influential in contribution to the place-making process for each of these Afghan women in the culturally diverse city of Auckland.

Public presentation of Islam in the West and New Zealand

Islam has generally been reported on negatively by Western media, which tends to conflate it with issues of national security and terrorism. Studies show that Western media has continued to report negative stories about Islam and Muslims, influenced by the news of terrorism, all around the world (Ahmad, 2006; Ahmed and Matthes, 2017; Ewart et al., 2016; Kabir and Bourk, 2012).

Media, the society's source of information, has shaped and continues to shape an image of Islam and Muslims which is not generally accurate. It is important then to acknowledge the religious composition of the Afghan population living in Auckland for the purpose of this research. According to Census

Table 9.1 Key characteristics of participants in the research

Name of participant	*Soraya*	*Banoo*	*Zahra*	*Meena*	*Sadjida*	*Taiba*	*Henna*	*Sophia*
Ethnicity	Hazara	Hazara	Hazara	Hazara	Hazara	Hazara	Hazara	Tajik
Religion	Shia Muslim	Shia Muslim	Shia Muslim	Shia Muslim	Shia Muslim	Shia Muslim	Shia Muslim	Shia Muslim
Age	45	44	21	18	23	22	25	25
Education	Illiterate	Elementary	Undergraduate studies	Intermediate school	Bachelor's in Psychology	Undergraduate studies	Undergraduate studies	Bachelor's in Sociology and Philosophy
Occupation – 1st phase	Unemployed	Unemployed	University student	School student	Teacher's aid in Epsom Girls' High School	University student/ Oral health assistant	University student	Unemployed
2nd Phase	–	–	–	Restaurant manager	Nurse	Assistant in a dental clinic	Receptionist	Unemployed (mother of two children)
Citizenship status in New Zealand	NZ Resident	NZ Resident	NZ Citizen	NZ Citizen	NZ Citizen	NZ Citizen	NZ Citizen	NZ Citizen
Time living in NZ – 1st phase	3 years	3 years	15 years	15 years	13 years	12 years	14 years	15 years

continued

Table 9.1 Continued

Name of participant	*Soraya*	*Banoo*	*Zahra*	*Meena*	*Sadjida*	*Taiba*	*Henna*	*Sophia*
City/country of birth	Kabul/ Afghanistan	Kabul/ Afghanistan	Tehran/Iran	Tehran/Iran	Peshawar/ Pakistan	Kabul/ Afghanistan	Kabul/ Afghanistan	Kabul/ Afghanistan
City/country of residence before migration to NZ	Mashhad/Iran	Quetta/ Pakistan	Tehran/Iran	Tehran/Iran	Peshawar/ Pakistan	Peshawar/ Pakistan	Peshawar/ Pakistan	Peshawar/ Pakistan
Age at time of migration from Afghanistan	6	2	Born in Iran	Born in Iran	Born in Pakistan	2	7	5
Age at time of immigration to New Zealand	42	41	6	3	10	10	11	11
Marital status – 1st phase	Married – Husband left the family (not divorced)	Married – to second husband. The first husband passed away in Pakistan	Married	Single	Single	Engaged	Single	Married
2nd phase	–	–	–	Single	Single	Married	Single	Married
Number of children – 1st phase	3	6	No children	0	0	0	0	1
2nd phase	–	–	–	0	0	0	0	2

2013, a majority of the Afghan population (96 per cent) in Auckland identified themselves as Muslims (Statistics New Zealand, 2013), while less than 1 per cent noted Christianity as their religion and less than 1 per cent mentioned that they had no religion. The interactive data from Statistics New Zealand does not provide an interaction between sex, religion and ethnicity. Therefore, an assumption is that a considerable proportion of Afghan women, relative to their total population in Auckland, are Muslim. This assumption results from the patriarchal structure of the family in Afghan culture and the way that Sharia Law (Sharia law (Arabic: شريعة) is the body of Islamic law) positions men as the leaders of the family and expects women to follow their men in different aspects of life including religion (Emadi, 2002; Esposito and DeLong-Bas, 2002). Also, the Afghan community presents itself as a Muslim community in Auckland. This is confirmed by their presence and participation as a whole community in Islamic events and celebrations in Auckland, including mourning sessions for Shia Muslims and Eid. I will expand on this further in the chapter.

Although the influence of religion is different from country to country, and from culture to culture, respect for people's freedoms of religion, including Islam, is enshrined in New Zealand's secular state and government. However, multicultural policy structure may deliver specific laws that prioritise one religion over others (Kolig, 2010). Therefore, a totalising view does not respond to those cross-cultural and intercultural differences in religion, ethnicity and culture. For example, in the case of refugees who identify themselves in the category of Middle Eastern or Asian, there are many diverse cultural and religious groups that need to be acknowledged by policy makers in terms of their differences as well as their similarities during their (re)identification process in their new living place.

Afghan women's sense of community in Auckland

Studies on community and social bonding suggest that in order to be a member of a community, people need to feel a sense of belonging to the common values of the community shared between all the members (McMillan and Chavis, 1986; Perkins and Long, 2002). Studies on Afghan people show that their association with the Muslim community comes from their sense of rootedness in Islam as the main aspect of their culture (Emadi, 2005; Johnson, 2004). Islam is 'the one component of identity that is shared by almost all Afghans, providing a common frame of reference for morality, rights, and obligation' (Johnson, 2004: 6). Additionally, members of a community gain a sense of belonging to their communities because they carry collective memories of their community-related experiences. Eventually, a sense of belonging and an inclination towards participating in those experiences make members feel affiliated to the community, whether it is territorial or relational,[4] or a combination of both (McMillan and Chavis, 1986; Perkins and Long, 2002). To clarify, a territorial community is a geographical and locational one, such as a neighbourhood, city or town. A relational community is one of human relationships without a specific geographical setting (McMillan and Chavis, 1986). However, these two notions of community are not exclusive. Territorial communities, such as neighbourhoods, have a potential to be relational, since people in a neighbourhood may develop their communities by sharing their interests and values (Gusfield, 1975; Sampson, 1988).

The findings from in-depth interviews with eight Afghan Muslim women in Auckland demonstrate that these women see themselves as part of a community, by referring directly to a specific community to which they felt a sense of affiliation and belonging. This confirms their sense of membership, which is a key to individuals developing a sense of community (McMillan and Chavis, 1986; Perkins and Long, 2002). Numbers of the younger generation of immigrants referred to the Afghan community, and more specifically the Hazara community, to which they feel affiliated. Therefore, they feel a sense of belonging to the common values of the Afghan/Hazara community in Auckland. However, the sense of belonging could include other communities. Some other participants from the 1.5 generation referred to other communities through which they share some common aspects of identity, while other aspects of their identity diverge. The 'Muslim community' is an example of another community to which numbers of the participants feel affiliated. On the other side, Afghan women who belonged to the first generation of immigrants expressed their sense of belonging to the Afghan/Hazara community in different ways. While, their past life experiences were traumatically directed and regulated by the patriarchal familial/ethnic/religious culture in Afghanistan, Pakistan and Iran before moving to New Zealand, the different narratives of their daily lives in Auckland confirmed their strong sense of attachment to the Shia Muslim culture as an ingrained part of their identity.

Investigations were also made as to how these women fitted into their new spatial (locational) settings, using both conventional and cognitive mapping techniques including participants producing sketches of their neighbourhood and the main places to which they related (see cognitive mapping, Tafaroji, 2017). Religion, as the absolute basis of Afghan collective culture, was found to be a key facilitator (and 'bridge') in how these women developed a sense of belonging in relation to particular places within the city, and how they negotiate and relate to the dominant New Zealand discourse. Religion is an imperative aspect of Afghan women's identity and has a strong influence on their everyday life experience. Consciously and unconsciously, most of the participants referred to the religious events and gatherings in Auckland, which mostly take place in the Pakuranga Hussainia, a religious place of worship for Shia Muslims in the suburb of Pakuranga. Gatherings of the Afghan community, though, are not exclusively religious occasions. There are gatherings of the Hazara community held weekly on beaches around Auckland (mainly at Long Bay Beach on the North Shore). As a result of experiencing the same place and the same people, participant Afghan women maintain their friendship ties and the social bonds between themselves and other members of their ethnic community. Except during the wintertime, most of these outdoor activities and gatherings take place at Long Bay Beach/Regional Park, which also provides an opportunity to be visible to other people (non-Afghanis) in a public place. So, while they are experiencing their community gathering and strengthening their social bonds within their community, they are also exposing themselves to other people who do not belong to the Afghan community. Therefore, there is always a chance of interacting with and encountering others in the public space while carrying on their community-related activities. There are other gatherings which are not associated with their ethnic or religious community. Participants from the 1.5 generation of immigrants referred to festivals and occasions like Christmas, Diwali and the Lantern Festival as events with which they feel a sense of familiarity. Diwali festival is the most popular event for participants in this study from both generations. Diwali festival, which is a Hindu festival run mainly by the City of Auckland, relates Afghan women to some aspects of their identity including their culture. The similarities between their culture and Indian culture provide an opportunity for the members of the Afghan/Hazara community to feel connected with the festival. This contributes to the formation and development of the geographies of (intercultural) encounter for Afghan women in their host society during their experience of co-existing with different individuals and communities in time and place (Friedman, 1998; Valentine, 2008; Wilson, 2017).

From the quotations of participants from the younger generation, I conclude that having shared experiences in place and time leads to the creation and development of images and meanings for them from that place. This could be discussed as the creation of transnational social spaces (Massey and Jess, 2003 [1995]), in which different people share their images of the place based on their different layers of identity in time. Additionally, through the process of (re)territorialisation of the place, this group of young Afghan women signify meanings, memories and identities of the place, which are shared between them. However, in the case of participants from the first generation of immigrants, a lack of expression of interest in the Afghan community, and their negative image of the wider Afghan community in Auckland, has resulted from the conflictive experiences they faced during their lifetime in the deterministic traditional and religious culture. They were consciously unwilling to repeat the same experiences in the host society of Auckland. They wanted to have a 'better life'. Nonetheless, religion, ethnicity and other rooted aspects of their identities connect them to the Afghan community in Auckland through community-related experiences.

The spatial implication of urban policies on Afghan Muslim women's right to the city

The spatial configuration of Auckland is also different from what these women were used to in their home countries. In towns and villages in Afghanistan, the locations of home and work were likely to be close, as is the case in most traditional close-knit Muslim urban areas. But, in Auckland, New Zealand (and in many Western cities) the home–work relationship is likely to be intentionally separated by town planning policies that have been imposed via Western land-use zoning in the city (as already explained in Mokarram's chapter). The zoning policies introduced by the planners to create 'well-ordered' and low-density cities, along with the growth of low-density suburban residential areas at greater and greater distances from the Central Business District (CBD), have resulted in highly car-oriented cities, as seen in the urban morphology of Auckland (Austin et al., 2014; Dixon and Dupuis, 2003; Memon, 1991; Saville-Smith, 2000). Such policies divided housing areas from working areas and reinforced the division of labour between men and women, between the male 'breadwinner' and the 'housewife' that stays at home (McDowell, 1997) and thus divided the city too (McDowell, 1999).

Therefore, suburbanisation and divisive zoning regulations have actually strengthened adherence to the traditional gender roles within the Afghan population living in Auckland. This is because it makes it so much more difficult for women to combine their traditional home-based roles with work outside the home which usually requires travelling to distant workplaces.

Conclusion

While in the last chapter Mokarram's research showed that male/female gender divisions within the Muslim city are not as 'absolute' as may be imagined but in fact that there is a fine grain of differences, layers and degrees of division, especially in the 'hidden' areas behind the main public squares and boulevards, this chapter has also demonstrated different place-making opportunities created by Muslim Afghan women through the intersection of different layers of their identity as individuals and within communities.

This chapter looked at the development of the social bonding of community members to the places where their collective experiences take place. The discussion shows that the participant Afghan women shape and develop geographies of encounter through crossing the boundaries and borders between groups and communities. These are territories which are (re)created through the intersection and encounter of different aspects of their identity (including their ethnicity, gender and religion as well as their stage of life) in Auckland. The cultural bonds between Afghan people, resulting from their collective culture, play an important role in their social bonding in Auckland.

Analysis in this chapter of the findings of the study also implies that there are considerable differences in the participants' perceptions of the community, their sense of affiliation and their images of places in the city. These differences shape different experiences in the place, which lead to the formation and development of the geographies of encounter (Ahmed, 2000; Amin, 2008; 2015; Carter, 2013; Friedman, 1998; 2001; Valentine, 2008; Wilson, 2017). The participants' different encounters would then lead them to different senses of community and social bonding in the bigger society.

To recap, the interaction of religion, ethnicity and gender, among many other layers of identity, shapes the nature of community development and social bonding.

References

Abu-Lughod, L. (2002) 'Do Muslim women really need saving? Anthropological reflections on cultural relativism and its others', *American Anthropologist* 104(3): 783–790.

Ahmad, F. (2006) 'British Muslim perceptions and opinions on news coverage of September 11', *Journal of Ethnic and Migration Studies* 32(6): 961–982.

Ahmed, S. (2000) *Strange Encounters: Embodied Others in Post-coloniality*, London and New York: Routledge.

Ahmed, S. and Matthes, J. (2017) 'Media representation of Muslims and Islam from 2000 to 2015: a meta-analysis', *International Communication Gazette* 79(3): 219–244. doi: 10.1177/1748048516656305.

Amin, A. (2008) 'Collective culture and urban public space', *City* 12(1): 5–24.

Amin, A. (2015) 'Animated space', *Public Culture* 27(2 (76)): 239–258.

Austin, P.M., Gurran, N. and Whitehead, C.M. (2014) 'Planning and affordable housing in Australia, New Zealand and England: common culture; different mechanisms', *Journal of Housing and the Built Environment* 29(3): 455–472.

Baker, A. (2010) 'Afghan women and the return of the Taliban'. *Time Magazine*, 9.

Barfield, T. (2010) *Afghanistan: A Cultural and Political History*, Princeton, NJ: Princeton University Press.

Beaglehole, A. (2013) *Refuge New Zealand: A Nation's Response to Refugees and Asylum Seekers*, Otago: Otago University Press.

Beebeejaun, Y. (2017) 'Gender, urban space, and the right to everyday life', *Journal of Urban Affairs* 39(3): 323–334. doi: 10.1080/07352166.2016.1255526.

Bhanji, R.M. (2011) *Resiliency amidst the Fragmented Lives of Afghan Refugee Women*. Master's dissertation, Open Access Dissertations and Thesis. http://hdl.handle.net/11375/11258.

Carter, P. (2013) *Meeting Place: The Human Encounter and the Challenge of Coexistence*, Minneapolis, MN: University of Minnesota Press.

Clements, F. (2003) *Conflict in Afghanistan: A Historical Encyclopaedia*, Santa Barbara, CA, Denver, CO and Oxford: ABC-CLIO.

Crenshaw, K. (2019) *On Intersectionality: Essential Writings*, New York: New Press.

Dixon, J. and Dupuis, A. (2003) 'Urban intensification in Auckland, New Zealand: a challenge for new urbanism', *Housing Studies* 18(3): 353–368. doi: 10.1080/02673030304239.

Dupree, L. (1980 [1973]) *Afghanistan*, Princeton, NJ: Princeton University Press.

Elliott, A. and Lemert, C. (2009) *The New Individualism: The Emotional Costs of Globalization*, London: Routledge (revised edition)

Emadi, H. (2002) *Repression, Resistance, and Women in Afghanistan*, Westport, CT: Praeger.

Emadi, H. (2005) *Culture and Customs of Afghanistan*, Westport, CT: Greenwood Press.

Eposito, J. and DeLong-Bas, N. (2002) *Women in Muslim Family Law*, Syracuse, NY: Syracuse University Press.

Ewart, J., Pearson, M. and Healy, G. (2016) 'Journalists' and educators' perspectives on news media reporting of Islam and Muslim communities in Australia and New Zealand', *Journal of Media and Religion* 15(3): 136–145. doi: 10.1080/15348423.2016.1209391.

Fenster, T. (1999) *Gender, Planning, and Human Rights*, London: Psychology Press.

Fraser, N. (1998) 'Social justice in the age of identity politics: redistribution, recognition and participation', *The Tanner Lectures on Human Values* 19: 72–91.

Friedman, S.S. (1998) *Mappings: Feminism and the Cultural Geographies of Encounter*, Princeton, NJ: Princeton University Press.

Friedman, S.S. (2001) 'Locational feminism: gender, cultural geographies, and geo political literacy', in M. Dekoven (ed.) *Feminist Locations: Global and Local, Theory and Practice*, New Brunswick, NJ and London: Rutgers University Press, 13–36.

Ghasemi, M.E. (1998) 'Islam, international human rights and women's equality: Afghan women under Taliban rule'. *Southern California Review of Law and Women's Studies* 8: 445–449.

Gilbertson, A. and Maeres, C. (2013) *Ethnicity and Migration in Auckland*. Technical Report No. TR2013/012. Auckland: Auckland Council. www.Aucklandcouncil.govt.nz/SiteCollectionDocuments/aboutcouncil/planspoliciespublications/technicalpublications/tr2013012ethnicityandmigrationinAuckland.pdf.

Gupta, A. and Ferguson, J. (1992) 'Beyond "culture": space, identity, and the politics of difference', *Cultural Anthropology* 7(1): 6–23.

Gupta, A. and Ferguson, J. (1997) *Culture, Power, Place: Explorations in Critical Anthropology*, Durham, NC: Duke University Press.

Gusfield, J.R. (1975) *Community: A Critical Response*, New York: Harper & Row.

Jarvis, H., Cloke, J. and Kantor, P. (2009) *Cities and Gender*, London: Routledge.

Johnson, C. (2004) *Afghanistan*, Oxford: Oxfam (2nd edition).

Kabir, S.N. and Bourk, M. (2012) 'Representing Islam and Muslims in New Zealand newspapers', *Journal of Muslim Minority Affairs* 32(3): 324–338. doi: 10.1080/13602004.2012.727293.

Kail, E. and Irschek, E. (2013) 'Vienna: progress towards a fair shared city' in I. Sánchez de Madariaga and M. Roberts (eds) *Fair Shared Cities*, London: Ashgate, 193–230.

Khan, A. (2002) 'Afghan refugee women's experience of conflict and disintegration', *Meridians* 3(1): 89–121.

Kolig, E. (2010) *New Zealand"s Muslims and Multiculturalism*, Leiden: Brill.

Lahav, G. (2003) 'Migration and security: the role of non-state actors and civil liberties in liberal democracies'. Paper presented at the *Second Coordination Meeting on International Migration*, United Nations, Department of Economic and Social Affairs, Population Division, New York.

Liev, M.H. (2008) *Adaptation of Cambodians in New Zealand: Achievement, Cultural Identity and Community Development*. PhD in Development Studies, Auckland: University of Auckland.

Lipson, J. G. and Miller, S. (1994) 'Changing roles of Afghan refugee women in the United States'. *Health Care for Women International* 15(3): 171–180.

Malkki, L. (1992) 'National geographic: the rooting of peoples and the territorialisation of national identity among scholars and refugees', *Cultural Anthropology* 7(1): 24–44.

Marlowe, J. (2015) 'Belonging and disaster recovery: refugee-background communities and the Canterbury earthquakes', *British Journal of Social Work* 45(suppl 1): i188–i204.

Marsden, P. (2001) *Afghanistan: Minorities, Conflict and the Search for Peace*, MRG Report, Minority Rights Group International.

Massey, D. and Jess, P.M. (eds) (2003 [1995]) *A Place in the World?: Places, Cultures and Globalization*, Oxford and New York: The Open University; Oxford University Press.

McDowell, L. (1997) *Capital Culture: Gender at Work in the City*. Malden, MA: Blackwell.

McDowell, L. (1999) *Gender, Identity and Place: Understanding Feminist Geographies*, Oxford: Polity Press.

McMillan, D.W. and Chavis, D.M. (1986) 'Sense of community: a definition and theory', *Journal of Community Psychology* 14(1): 6–23. doi: 10.1002/1520-6629(198601)14:13.0.CO;2-I.

Memon, P.A. (1991) 'Shaking off a colonial legacy? Town and country planning in New Zealand, 1870s to 1980s', *Planning Perspective* 6(1): 19–32.

Mohammad, R. (2005) 'Geographies of Muslim women: gender, religion, and space', in G. Falah, and C. Nagel (eds) *Geographies of Muslim Women: Gender, Religion, and Space*, New York: Guilford Press, 178–200.

Mohanty, C.T. (1988) 'Under Western eyes: feminist scholarship and colonial discourses', *Feminist Review* (30): 61–88.

Nicholson, L.J. (2008) *Identity before Identity Politics*, Cambridge and New York: Cambridge University Press.

Perkins, D.D. and Long, D.A. (2002) 'Neighbourhood sense of community and social capital', A.T. Fisher, C.C. Sonn and B.J. Bishop (eds) *Psychological Sense of Community: Research, Applications and Implications*, New York: Springer, 291–318.

Ross-Sheriff, F. (2006) 'Afghan women in exile and repatriation passive victims or social actors?', *Affilia* 21(2): 206–219.

Rostami-Povey, E. (2007) *Afghan Women: Identity and Invasion*, London: Zed Books.

Said, E.W. (1993) *Culture and Imperialism*, New York: Knopf.

Said, E.W. (1997) *Covering Islam: How the Media and the Experts Determine How We See the Rest of the World*, London: Vintage Books (revised edition).

Sampson, R.J. (1988) 'Local friendship ties and community attachment in mass society: a multilevel systemic model', *American Sociological Review* 53(5): 766–779.

Saville-Smith, K. (2000) 'Positioning women and children in New Zealand's urban life', in P.A. Memon and H.C. Perkins (eds) *Environmental Planning and Management in New Zealand*, Palmerston North, NZ: Dunmore Press, 339–347.

Spoonley, P. and Butcher, A. (2009) 'Reporting superdiversity: the mass media and immigration in New Zealand', *Journal of Intercultural Studies* 30(4): 355–372.

Statistics New Zealand (2013) *2013 Census. Ethnic Group Profile: Afghani*. www.stats.govt.nz/Census/2013-census/profile-and-summary-reports/ethnic-profiles.aspx?request_value=24755&tabname=Keyfacts.

Statistics New Zealand (2014) *2013 Census. QuickStats about Culture and Identity*, Wellington: Statistics New Zealand. www.stats.govt.nz.

Tafaroji, R. (2014) '(Re)territorialization as a daily process for Afghan immigrant women in Auckland', Presented paper at NZGS (New Zealand Geographical Society) Conference, Auckland, New Zealand.

Tafaroji, R. (2017) 'Analyzing a global sense of place by using cognitive maps: a study of Afghan immigrant women in Auckland', *Spaces of Dialog for Places of Dignity*, AESOP Conference, Lisbon: Aesop Conference Proceedings, ID Number 1420, pp. 1460–1473.

Tafaroji, R. and Reeves, D. (2014) 'The everyday life of immigrant women as a struggle to create their own living places,' Paper presented at *Designing Sustainable and Inclusive Urban Environments*, Gender STE Engendering Cities, Rome, Italy, 24–25 September 2014.

Thorns, D.C., Fairbairn-Dunlop, T.P. and Du Plessis, R. (2010) 'Biculturalism, cultural diversity and globalisation: issues for Aotearoa New Zealand', *Korean Social Science*.

Valentine, G. (2008) 'Living with difference: reflections on geographies of encounter', *Progress in Human Geography* 32(3): 323–337.

Ward, C. and Masgoret, A. (2008) 'Attitudes toward immigrants, immigration, and multiculturalism in New Zealand: a social psychological analysis', *International Migration Review* 42(1): 227–248.

Wilson, H.F. (2017) 'On geography and encounter: bodies, borders, and difference', *Progress in Human Geography* 41(4): 451–471.

Woodward, I., Skrbis, Z. and Bean, C. (2008) Attitudes towards globalization and cosmopolitanism: cultural diversity, personal consumption and the national economy', *The British Journal of Sociology* 59(2): 207–226.

Young, I.M. (1986) 'The ideal of community and the politics of difference', *Social Theory and Practice* 12(1): 1–26.

Young, I.M. (2000) *Inclusion and Democracy*, Oxford and New York: Oxford University Press.

Young, I.M. (2011 [1990]) *Justice and the Politics of Difference*, Princeton, NJ: Princeton University Press (Paperback reissue with a new foreword by Danielle Allen).

Part IV

Non-religious space and women

10 Postsecular urbanism and the hidden value of gendered spaces

Katie McClymont

Abstract

This chapter explores issues of religion and spirituality beyond organised faith systems and designated places of worship, and how this affects gender and urban life. It explores the gender implications of the concept of 'municipal spirituality': a notion that local government (and hence planners) both could and should promote and defend places whose value is intangibly spiritual, through shared belonging, or places which offer some sorts of connection beyond the material. These include many public assets such as libraries, parks, community centres and cemeteries, places whose value is not only the instrumental service they provide. Urban planning has a poor track record in dealing with issues of religion and of gender. Professional planning very much conceives of itself as a secular, rational and objective practice. However, as this paper argues, this can make the articulation of the value of places of 'municipal spirituality' very difficult, giving planning limited grounds to develop or preserve such spaces. The gender implications of this are then explored, drawing on both classic and contemporary analyses of how urban space is gendered, considering ways that 'public', 'private' and 'community' are conceived, and how this devalues and renders invisible both more typically female experiences of everyday life and deeply held emotional and spiritual attachment to place.

Key words: postsecular, community, municipal spirituality, cemeteries, women

Introduction: spaces of faith, gender and community

This chapter explores the gendered dimensions of spiritual experience and values that fall outside the mainstream and commonly accepted understandings of 'religion', and how this relates to issues of planning and urban studies. Specifically, it considers how the management of urban space and vocabularies of planning currently occludes and obscures 'implicit religion' (Bailey, 1997; 2010) and subsequently the existence and importance of spaces of 'municipal spirituality' (McClymont, 2015; 2018). These spaces include – but are not limited to – community centres of all descriptions (old, modern, purpose-built, re-purposed, temporary, longstanding), public parks and cemeteries, some public leisure facilities including swimming pools, and libraries. The importance of these places, and the difficulties in both defining them and clearly articulating their value in planning terms, is discussed in greater depth later in this chapter. The chapter argues that the limited understanding of such spaces is intrinsically linked to the gendering of society and how this in turn impacts on the spatial makeup of cities and towns.

Planning and public policy more broadly situate their underpinning rationale within a secular framing, in which religion is something private, discretely categorised and tightly bounded within designated 'places of worship'. This framing excludes the possibility and importance of versions of spirituality or religious belief and values which do not fit this secular compartmentalisation. This situation is then considered through a lens of gender, drawing on arguments of Joan Wallach Scott (2017) and others (Aune et al., 2017; Braidotti, 2008) which problematise assumptions about secularism, religion and women. This demonstrates more fully the difficulties inherent in the way secular policy frameworks deal with issues of religion. By questioning the position and maintenance of the mainstream divide between the sacred and secular, and seeing how (implicit) religious values permeate a wider range of places and aspects of daily life, often concomitant with gendered notions of 'caringscapes' (Bowlby et al., 2010), the argument here further challenges the construction of religion as something private and individualised. The current divisions in our policy understandings, and their underpinning social ontology, unhelpfully divide notions of public/private and sacred/secular. In turn this has implications on the lived and built reality of twenty-first-century places and how 'community' spaces, valued and experienced as part of the daily life of many women, are eclipsed from social importance.

This is of pressing importance primarily in a UK context (Robinson and Sheldon, 2018). The community empowerment campaigning and support organisation Locality, in its 2018 report *The Great British Sell Off*, highlights how over 4,000 publically owned assets are being sold off by local authorities each year, including youth centres, libraries, publically owned buildings in parks and swimming pools. Although not all of these assets may have value as expressed by the term 'municipal spirituality', this scale of change is worrying. As well as demonstrating the quantitative scale of these losses, the report states 'that the deeper value to local people will never be realised' (Locality, 2018: 5), hinting at a sense that these places hold a deeper importance than just that of the function they fulfil. However, the *gendered* nature of the use of such spaces is not brought to the fore: such as the use of church halls for playgroups predominantly organised for and by young mothers, something which seems to be 'below the radar' of many planners. It is widely under-examined despite claims being made about the importance of ensuring public spaces are suitable for a wide range of groups, including families with young children and older people, as they 'create a wider sense of attachment and discovery' (Worpole and Knox, 2007: 14). It is this agenda that the chapter aims to take forward. Specifically, in challenging the assumptions about public/private/sacred/secular divides, the chapter both identifies and re-values spaces which enable spiritual connections for those who are not part of an organised 'world' religion. Consequently, this enables a clearer understanding of the importance of certain 'community' buildings and spaces, highlighting overlooked aspects of (predominantly women's) everyday life.

These arguments are set within a context of postsecularism (McLennan, 2010; 2011) and specifically postsecular cities (Baker and Beaumont, 2011; Baker, 2016; Baker and Dinham, 2017), as introduced by Chris Baker in Chapter 2. Postsecular here does not mean a simple end to secularism, or to the historical and social processes wherein religion plays a less important role both in the running of societies and in creating and enforcing the 'rules' which guide individual daily life. What the concept does draw attention to is the complexity and diversity of contemporary religious experience. It notes how secularisation has been uneven, contextually specific and at times an inaccurate description of social changes, especially when looking beyond Western Europe, but also noting the increase in mosques and majority Black Pentecostal mega churches in the UK and Europe, and informal 'roadside' memorials (Petersson, 2010; Greed, 2016a; Gale, 2008). Moreover, such understandings are fundamentally tied to notions of citizenship, which are problematically underpinned by ideas of white male individualism, which have had a troubled history of dealing with notions of rights of women (Scott, 2017; 1996; also see Greed, 2011 with specific reference to cities/urban planning). However, the concept is important in framing the arguments put forward here. The idea that there is a clear, established and universally agreed upon distinction between 'religious' and 'secular' is challenged, supporting claims that such divides do little other than reveal the views of the powerful, and in turn put (certain types of established) 'religion' in a pre-defined, privatised box, something which functions much like a hobby for certain people who choose to take it up. Moreover, such divides help to re-inforce unproblematised notions of public and private, and of individual freedoms and citizenship which are fundamentally – often unacknowledgedly – entangled with repressive and problematic ideas of gender and the relationship between men and women (Scott, 2017; Jupp, 2014). However, these assumptions in turn shape the understandings of space on which decisions in urban planning are made, decisions which prioritise certain building uses – such as those with clear economic functions (e.g. retail) or vital to tangible aspects of social reproduction (e.g. housing). Subsequently this re-enforces relationships and patterns of (gendered) identity and encounter with the value of (predominantly female) community space being at best second class and at worst, entirely invisible.

The chapter is structured in four main parts. It next discusses the idea of 'municipal spirituality' at greater depth, demonstrating how this relates to both implicit religion and to theological understandings of community and city space, as well as showcasing the sorts of spaces this definition can relate to and how and why this is important for women. It then draws upon discussions about gender, women and the idea of secularism, arguing how a broader understanding of religion, situated within the postsecular context outlined above, offers a more radical, pro-feminist perspective than mainstream critiques of women's status in many world religions would otherwise allow. Next, it discusses how women and gender issues have been dealt with in urban studies and planning literature, with particular attention to their role in the idea of 'community' and how this can unsettle notions of public and private in policy and activism. It then brings these ideas into discussion, demonstrating how the idea of municipal spirituality can aid feminist challenges to dominant conceptions of city space which underpin contemporary urban planning. In so doing, the value of community spaces can be seen in a new way. At the same time, this demonstrates how a broader understanding of spiritual values, and their relationship with spaces and places both brings new questions about value to the fore and offers possibilities for re-imagining the city.

Municipal spirituality (implicit sacred space)

The argument in this chapter relies on an understanding of the idea of municipal spirituality which I have developed and discussed in greater depth elsewhere (McClymont, 2015; 2018). However, it is necessary to introduce the idea here for the purpose of outlining its implications on thinking about planning, gender and religion in contemporary spaces and places. Municipal spirituality is a way of defining the qualities of a place which although not functioning explicitly as a place of worship, offers access to communal, shared belonging and to that which is beyond everyday material existence. These qualities are framed on the basis of an ontology which challenges both the logic and implications of instrumental, rational modernity (see McClymont, 2015: 540–542 for a fuller explanation of this). It is a position which compliments the post-foundationalist critiques of planning which challenge the hegemony of fixing meanings and understandings of the social (Wilson and Swyngedouw, 2015; Metzger et al., 2015), and links to the criticisms of 'disenchantment' discussed below. This is not merely an academic or theological issue, however. It relates directly to the use and function of lived spaces and places. In contrast to the majority of ways places are defined and understood within planning policy and practice, municipal spirituality draws out the *intangible* values of a place. In this way, different from 'leisure', 'housing' or 'employment' for example, whose functions are more readily explainable, I argue it is as necessary for a functioning community, and a vital aspect of what is most readily termed 'wellbeing' within planning circles.

However, because planning's vocabularies (and standard ontology) focuses on the tangible, functional aspects of a place, the religious or spiritual aspects of experience and life, not readily fitting these categories, cannot be clearly articulated, especially those which fall beyond mainstream categories of organised religion. Planning can provide for explicit 'places of worship'; albeit problematically, and with set ideas and understandings of size, design, scale and parking which often impede groups/faiths that are not majority white British (Gale and Naylor, 2002; Gale, 2008; Greed, 2016a; 2016b; Beebeejaun, 2004; 2012). This is because the conceptual category of 'a place of worship' is fundamentally secular; by default a designated place of worship means that all other places are not for worship, denying the spiritual value of places outside this category. In so doing, religion is privatised and commodified. It becomes spatially contained within discrete codified entities; a place of worship is a place for worshippers; specific, self-identified individuals who have, as pseudo-consumers, made an explicit choice to go there, and the space exists discretely to fulfil these specific needs. The chapter is not claiming that this is how many, any or most people who attend places of worship view their actions and beliefs. Arguments from some Christian theologians challenge the secularist ontology which frames places of worship in this way (see for example, Milbank, 2006; Cavanaugh, 1999) and the literature on 'lived religion' challenges the importance of formal boundaries for many people's experience of faith and belief (Primiano, 1995). The critique contained within this discussion of 'places of worship' is levelled at 'the state', policy and its underpinning secularist beliefs. This relies on categorical definitions of religion, the boundaries of which are fixed and policed by those with established power. In this, 'places of worship' have a distinctly different value to the sorts of community assets mentioned above as places of municipal spirituality. The postsecular context challenges such clear delineations; this can also be seen through the idea of 'implicit religion', which is discussed next.

Implicit religion, through the work and legacy of Edward Bailey, provides a useful counterpoint to many of the dominant assumptions about the category of religion. Bailey's work (see for example 1997; 2010), and the work of those drawing on this idea, aims to investigate phenomena and behaviour that are beyond the standard scope of definitions of religion, but when looked at in detail fulfil many of the assumed criteria that define religious practices. It is a way of looking at practices and behaviours such as environmentalism (Bartkowski and Swearingen, 1997; Dunlap, 2006) which critically interrogates our understandings of both 'religion' and 'secular' behaviour. In this way, it is akin to the idea of municipal spirituality – with places of municipal spirituality representing a sort of 'implicit sacred space' (McClymont, 2018). By challenging these categories in this way, it is possible to re-evaluate both the value and the role of a building or space in a way which is not common in planning practice.

Cemeteries: planning for death

To explain these ideas further, it is necessary to ask what this means in everyday practice; what are the implicit sacred spaces, or spaces of municipal spirituality in any town or city? In many ways, it is both impossible and contradictory to over-define this as its key defining quality is that of intangibility. In my previous work (McClymont, 2015; 2016), I have discussed how public cemeteries can be seen in this way: spaces which are simultaneously civic (owned by a local authority) and sacred (places where many people think of afterlife/prayer in more or less formally religious ways). In this, they are also very public, with access for all, similar to a public park, and very private and intimate, where people express memorial practices in diverse, individual and at times conflicting ways (Maddrell, 2011; Woodthorpe, 2011). As social, communal spaces, they offer something which is inclusive and open, yet meaningful and emotionally charged. This sort of space can therefore challenge the standard notions of religion, and the boundaries between sacred and secular space. In the UK, planning has very limited engagement with cemetery provision, with under 20 per cent of local authorities providing 'active' policies for the provision of cemeteries or crematoria (McClymont, 2014). This is problematic beyond just the possibility of running out of space for burial – although that is of great importance in itself! If not properly planned for, the location of such spaces may be inaccessible in terms of public transport, but also located in sites which feel isolated and unsafe, limiting the possibility of them offering shared space – places where people go not solely for funerals or to visit a specific grave. At best, cemeteries offer the space for cross-cultural connections, as well as quiet reflection and dialogue beyond everyday material existence. Although this is by no means exclusive to women, gender differentials in life expectancy mean more female partners will be tending the grave of a deceased husband/partner. More widely, gendered patterns of working/caring, as discussed below, lead to women spending more time in the activities outside paid employment and hence using more spaces which are not simply 'home' or 'work'. The categories of 'parks and greenspace' or 'green infrastructure', which most cemeteries in the UK are classed as, does not adequately promote or protect what is special about contemporary cemeteries and the lack of policy engagement here belies the lack of consideration to this.

Moreover, the idea of municipal spirituality provides a useful way to think about the value of community assets including libraries, pubs and community centres. In all of these examples, places of municipal spirituality are also 'community' spaces. Community centres and church halls provide places for people to meet outside the usual commercially centred spaces such as cafés or commercial leisure centres, which charge for activities. These are vital aspects of daily life for people – still predominantly women – who are caring for preschool children, as well as for older people (ONS, 2016). With patterns of daily life which do not revolve around the spaces and timescales of paid work, community space, often located outside main town or city centres, provides places for meaningful connections to be made. This is partially acknowledged in policy discourses – 'public spaces form a shared spatial resource from which experiences and value are created in ways that are not possible in our private lives alone' (Mean and Tims, 2005, cited in Worpole and Knox, 2007: 7) – but the vocabulary is still lacking and underdeveloped. The sorts of connections provided within such places, and the value they in turn provide to those who experience and co-create them, do not simply or solely provide a material function – cheap tea or coffee and an hour out of the rain and/or the house with a noisy two-year old are important and valuable in themselves but the ongoing connections and senses of belonging that such places can provide cannot only be defined in such tangible terms. It is their practical functionality – a large space, hireable at affordable/not-for-profit rates, well located near public transport routes and/or other facilities such as schools and medical centres, with storage cupboards for play equipment – which enables their intangible value to develop. Without planning providing for or protecting such spaces the connections which develop within them cannot take root, removing a valuable aspect of shared common life because the place for its enactment was lost. This area remains substantially under-researched, however, which in itself belies the limited importance given to questions of daily life and the use and value of space.

Here, the connection with questions about gender and women's experiences begins to align with questions of spirituality and spaces. This is considered in greater depth in the third and fourth sections of this chapter with parallels between gendered understandings of 'care' (Jupp, 2014; Bowlby et al., 2010; Gilroy and Booth, 1999) and places of municipal spirituality having overlapping and complimentary concerns. However, the chapter next turns to address questions of secularism and gender to conceptually group the importance of the arguments being made here.

Women/gender and secularism

This section outlines some of the debates around secularism and women. In particular, it draws on the work of philosopher and historian Joan Wallach Scott (2017) from her book *Sex and Secularism*. Her argument pivots on the idea that claims of women's rights being a fundamental aspect of the liberal secular state are, at best, oversimplifications. She controversially argues that '(g)ender inequality is not simply the by-product of the emergence of modern western nations … inequality is at its very heart. And secularism is the discourse that has served to account for this fact' (2017: 4). This section outlines the central tenets of her argument to demonstrate how this problematisation of gender equality and secularism can then challenge established norms and values within planning and city management, specifically with relation to ideas of 'public' and 'private' space. To do this, it then draws upon recent and classic debates in planning and urban scholarship about gendered assumptions and the role of women. In turn, this demonstrates how 'community' spaces – potential spaces of municipal spirituality – offer access to values and experiences which are occluded or denied by (planning's) secular ontology, values of implicit religion or undefined spirituality. In so doing, the chapter argues how this is intrinsically linked to ideas of gender and assumptions about men and women's roles.

Scott (2017) traces historic debates in the US and France, as well as Europe more widely, to demonstrate how, when societies became 'enlightened', establishing democratic governments and placing the authority of science above that of the church, this did not lead to claims of equality between men and women. Instead of rejecting differences between the sexes which stemmed from historical religious practice or Biblically based claims, enlightened men turned to science to ground arguments for gender difference (and inequality). Women's brains were argued to be underdeveloped compared to their male counterparts, to allow their reproductive functions to flourish. Moreover, the focus on women's reproductive functions became of renewed importance in times of what Weber classifies 'disenchantment'. To put it simply, the removal of a pervading belief in a god or an afterlife left society searching for meaning, and especially unable to deal with the idea of death. Securing descendants then became a necessity; the only way to 'live on' was via one's children, putting a renewed emphasis on both women's biological reproductive function, but also on their caring and nurturing nature, necessary to secure the future of the human race. Scott (2017) traces how this 'natural' divide is used to reinforce men's 'rationality' and therefore suitability for democratic governance, citizenship and equality, while at the same time necessarily denying this to women who remained relegated to pre-enlightenment caring roles.

The point of presenting her argument here is to demonstrate that the relationship between gender, (in)equality and secularism is more complex than it might otherwise appear. There is no more axiomatic link between religion and gender inequality than there is between secularism and gender inequality. These arguments are furthered by contemporary postsecular feminist thought (Braidotti, 2008). Claims for more humane, nature-respecting ways of life, alongside more established Black and post-colonial feminist arguments, challenge the dominance of rational reasoning and its coterminous disenchantment. Braidotti (2008) argues that '(t)he residual spirituality of much contemporary feminist theory demonstrates the compatibility of political subjectivity with issues that do not fall easily within the boundaries of the secular tradition in feminism' (Braidotti, 2008: 14). This can therefore be seen as challenging the established divide between secular and religious, and challenging the dominant claims about the role of women in both these realms. In so doing, it creates 'inclusive alternatives to humanist individualism and uncritical secularism' (Braidotti, 2008: 8). By unsettling an assumption that secularism is necessarily better for women than religious framings of society, these arguments challenge many taken-for-granted assumptions. First, they open up space for the voices of religious women (Aune et al., 2017) to be heard as a legitimate part of debate. Empirically, feminist scholarship has often neglected religious women (leading to their marginalisation), while religious inclusion scholars have neglected women (assuming that religious freedom is available to men and women equally) (Aune et al., 2017). Second, and with more wide-ranging consequences, they challenge the emphasis on rational instrumental (and subsequently capitalist) logics which exude from secular ontologies.

This discussion has revealed two important points for the purpose of this chapter. First, it has problematised the notion that a secular liberal order is necessarily better for women than one in which religion has some sort of public role. This includes drawing attention to the religious beliefs, experiences and lives of many women as valid and meaningful, as well as being more critically attentive to the gendered politics of secularism. Second, it has begun to highlight the problem of 'disenchantment' within secular societies: the ongoing search for some greater meaning, or ultimate reality which enables human existence to deal with the finitude of death. This is important because the values and beliefs displayed in places of municipal spirituality do just this; connect to a 'something' which is beyond material existence, or individual life. This is seen in the claims about the importance of community assets as detailed in McClymont (2015: 547–548, Boxes 1 and 2 in particular), which shows how people value shared spaces, here libraries and pubs, as the 'soul of a place'. Instead of women's reproductive functions replacing a god or sense of afterlife in a secular society, postsecular spaces of municipal spirituality tentatively suggest alternative means of 're-enchantment' without necessitating claims of natural gendered roles or a return to only an explicit codified notion of religious authority.

Public space, community and gender: women and planning

This section does not aim to rehearse the arguments made about gender in planning in any depth; this has been done adequately elsewhere, both in the introduction to this volume and elsewhere (for example Sánchez de Madariaga and Neuman, 2016; Beebeejaun, 2017). However, two points are worth raising and discussing in greater depth for the flow of this argument. These are related: first, the divisions between public and private, and second the linkage of women to community, as opposed to more civic, public roles which are seen as the role of men (Beebeejaun and Grimshaw, 2011; Jupp, 2014; 2017). In her classic paper of 1983, Linda MacDowell states that 'the division of urban space both reflects and influences the sexual divisions of labour, women's role in the family and the separation of home life from work' (MacDowell, 1983: 62). This is fundamental to a gendered understanding of urban, planned space. Cities are conceived, designed and built based on a patriarchal model of society in which the spheres of public/civic/work and private/domestic/home are both actively differentiated and spatially separate. Despite considerations about the oppressive nature of domestic work for women in key wartime reports such as Reith and Beveridge, and experiments in communal kitchens seen in Letchworth Garden City in its early inception (McDowell, 1983), the large-scale post-war housebuilding and New Towns programme did not further these radical ideas. Instead housing, both public and private, reinforced this notion of a gendered division of labour, especially within the suburbs. In parallel with the secularist paradigm outlined by Scott (2017), women's problematic, or second-class, status comes from their being aligned with the 'private/domestic/home' – and religious sphere. True citizenship comes only in the public sphere, made in an image of male rational autonomy, autonomy from domestic binds, and from the influence and emotional sway of religion. In turn, the way cities are organised reflects and compounds this inequality by maintaining this divide. Transport routes are designed to serve non-domestic working patterns, whereas housing, especially for families, is often seen as unsuitable in city centre locations, despite these having a greater provision of services. As is evidenced across Europe (Sánchez de Madariaga and Neuman, 2016; Altintas and Sullivan, 2016), women spend more time in domestic tasks, therefore are spatially situated within domestic spaces, making their experience and daily life of a lesser status, being 'neglected as political sites' (Beebeejaun, 2017: 330), a secondary service function to rational economic 'man'. The implications of this for questions of religion, and in particular for the idea of municipal spirituality and its relationship to gender, are discussed further below.

It is now necessary to turn to the second issue noted above: the relationship between women and 'community' functions. Women are traditionally seen as more suited to roles within community management and support, rather than politics and civic leadership, with such activities positioned as a natural extension of their domestic roles (Beebeejaun and Grimshaw, 2011; Jupp, 2014). Scott argues that this emerges out of the position women are placed into within secularism: 'the identification of women with the social, even as it eventually won them the right to vote, did not dispute the naturalised explanation for their difference and disqualifications and discriminations that followed from it' (Scott, 2017: 119). More directly related to questions of planning and urban governance, Beebeejaun and Grimshaw (2011: 1999) argue: 'Women are seen as having different ways of participating than men and, again, this is attributed to their gendered roles which place women in more informal processes contrasted to men who prefer more formal methods of organising'. However, what this means for (the value of) the different sorts of places where these activities take place – community and church halls, school annexes, libraries – remains under-researched.

This understanding, and valuing, of women's role in community spaces can be framed within debates about 'infrastructure of care' (Jupp, 2014: 1301; Bowlby et al., 2010; Gilroy and Booth, 1999). This work draws attention to the importance of everyday life, small-scale, locally based interventions in sustaining and promoting meaningful relationships and encounters which are outside the spheres of 'home' or 'public/civic life'. By being beyond and within both spheres, they in turn challenge this binary divide and its attendant problematic gender divisions: '(t)hrough the everyday spheres of life, the gendered and patriarchal nature of cities is more readily revealed and the agency of city dwellers operates to challenge it' (Beebeejaun, 2017: 330)

This operates on a number of spatial scales – challenging the boundaries of how planning perceives these places. As many community activities are both socially and spatially located in the domestic realm, they are found in the 'suburbs' and neighbourhood-based locations. However, community buildings such as halls, public libraries, local leisure facilities, cemeteries, parks and former or current church buildings cannot accurately be described as domestic/home spaces. Moreover, they do not fit standard ideas of public/civic/work, and their role is not part of either formal governance or commerce. These semi-public places (Peterson, 2017) do not neatly fit the public/private, male/female binary, and maybe that is why they are so overlooked and poorly theorised in much planning literature. This parallels the notion of implicit religion which underpins ideas of place of municipal spirituality. If we can step beyond the standard dualisms of public/private, secular/sacred, our interpretation and subsequent valuing of spaces and place can be radically altered. This can be done in a way which also challenges the negative normative assumptions and underlying power relations of gender dynamics.

Gender, planning and municipal spirituality: is the secular city bad for women?

The above sections demonstrate how notions of secularism and gender have largely been over-simplified, with assumptions about religion and women being made on the basis of liberal norms which do not pay sufficient attention to either historical or empirical detail of policy, discourse and experience. They demonstrate that much of what is claimed about gender equality in secular societies lacks a nuanced understanding of women's religiosity (Aune et al., 2017; Reilly, 2017) and is often complicit in orientalist constructions of 'non-Western' practices. Moreover, Scott's (2017) account, in particular, unsettles standard divides between what is 'religious' and what is 'secular', demonstrating that such categories are neither neutral nor fixed. The idea of implicit religion usefully frames some experiences and beliefs which the standard categorical divides occlude. This is where the idea of municipal spirituality – a category which in itself aims to transgress this polarity – can be useful to link these debates to everyday urban life and planning practice. This section now draws on the lens of gender, to interrogate further the meaning of municipal spirituality. It asks questions about how implicit religion, or implicit sacred spaces, play a role in shaping the city and what the gender implications and issues of this might be.

Places of municipal spirituality are likely to be predominantly female 'community' space (Jupp, 2014), space which challenges or compensates for gendered norms of work/domestic space (McDowell, 1983). Patterns of life outside private homes, public workplaces and places of consumption and leisure are largely ignored by planning policy, but as has been demonstrated above, central to many people's (especially women) lives. Such spaces offer the possibility for connections with others to be made which draw upon shared values, a sense of history and belonging, and links to something beyond everyday material existence. They demonstrate the possibilities of remaking shared values and shared meanings which do not only rely on rational logics of capitalism. Secularism denies this communal, public role of a sort of spirituality. It relegates it to the sphere of private consumption, and therefore leaves the provision of such spaces to the market. Municipal spirituality is a fundamentally public concept – public here to be understood as the antithesis of private; civic, communal or shared are also terms which can be used to describe its nature. As well as demonstrating the possibility of a different, implicit, postsecular understanding of religion, it offers the possibility to help change the contemporary political and ethical realms of debate, demonstrating 'the importance of community as a space to enter politics and enact social change' (Jupp, 2014: 1312) – a space of particular value to and for women.

This term can help to substantiate the value of women's everyday life as equally meaningful. By seeing spaces such as libraries, community halls, allotments – even informally, corners of roads where different paths cross and people have time and space to linger – as offering something fundamental to the core of human existence and wellbeing, municipal spirituality begins to offer a language to help re-value urban space. It can challenge the binaries of public/private, secular/sacred in a way which both challenges the gender roles on which they were established, and offers an alternative to the disenchantment of urban modernity. To understand this at greater depth, and to understand how this may vary between contexts and countries, further detailed empirical research is needed.

Conclusion

The idea of municipal spirituality demonstrates the implicitly religious values of places which are not categorised as 'places of worship', or sites of importance to a world religion. This poses a challenge to the dominant liberal, secular world view inherent within planning policy and planning judgements (Sandercock, 1998; Sandercock and Senbel, 2011). Moreover, this argument holds particular gendered implications. Spaces whose value is not readily recognised, articulated or codified by planning categories can hold central importance in the daily lives of (usually, but not exclusively) women (Locality, 2018; Robinson and Sheldon, 2018). They are places which offer deeper connections, both to other people within that space, and to something 'indescribable', potentially a notion of ultimate reality or that which is beyond material existence – a postsecular sense of re-enchantment. By defining and articulating the implicitly religious space wherein spiritual value is not over-codified or exclusive, municipal spirituality offers something beyond the standard binaries of secular/sacred, public/private. By validating religious or spiritual experience which is outside the domain of the 'world religions', municipal spirituality creates a vocabulary which can begin to define its role in cities and towns, and hence in urban planning, making a case for the importance of places which support, promote and provide for such interactions and shared values.

Moreover, these are places which can challenge the patriarchal underpinnings of urban space and citizenship, encapsulated in notions of 'public' and 'private' space, and their relative importance and acceptable functions in city planning. Transgressing this, places such as libraries, community centres and cemeteries can be both intimate and open; they are not for commerce, or direct democratic function. They can allow for greater connected social meaning than the private space of the home, or the rational disenchanted bureaucracy of the modern state. Concomitantly, a gendered lens of analysis helps to explain why these spaces are overlooked and poorly understood within mainstream planning and helps to suggest the new political questions which need asking if positive change can come about (Braidotti, 2008). They are placed in the 'female' realm, but at the same time they deny the simplistic binary on which this is founded. By highlighting their importance and value for the holistic human (and arguably non-human, but that argument is beyond the scope of this chapter), new questions can be posed about the type of society which we, as planners, are aiming to build. Feminist analysis of urban studies is not only to develop a gendered critique, but to invoke new normative visions of cities (Jupp, 2014).

Moreover, challenging a simplistic idea that secularism is good for women, with its attendant spatial implications that religion is something private/ised, municipal spirituality offers a new way for planning to embrace the challenges of postsecularism. It promotes spaces for new ways of belonging, and public engagement with deeper values and meaning than is usual within the acceptable scope of liberal (secular) democracies. By defending the importance of such places, municipal spirituality shifts the contours of political debate by refusing to accept established binaries of public/private, secular/religious, and the way that gender roles are made and remade within these. By suggesting that privatising and residualising religion does not simply 'free' women from 'traditional' constraints, Scott (2017) draws attention to the shifting political and ontological power structures of patriarchal societies. Questioning the simplicity of some secularist claims about gender and women's rights not only empowers religious women to be able to be in control of their own identity and beliefs, it also allows for a more subtle, implicit sense of religiosity to have resonance, meaning and value in urban planning and space.

References

Altintas, E. and Sullivan, O. (2016) 'Descriptive finding fifty years of change updated: cross-national gender convergence in housework', *Demographic Research* 35(16): 455–470. doi: 10.4054/DemRes.2016.35.16.

Aune, K., Lövheim, M., Giorgi, A., Toldy, T. and Utriainen, T. (2017) 'Is secularism bad for women?: La laïcité nuit-elle aux femmes?', *Social Compass* 64(4): 449–480.

Bailey, E. (1997) *Implicit Religion in Contemporary Society*, Den Haag: Kok Pharos.

Bailey, E. (2010) 'Implicit religion', *Religion* 40(4): 271–278.

Baker, C. (2016) 'Sustainable governance in a postsecular public sphere: re assessing the role of religion as a cosmopolitan policy actor in a diverse and globalized age', *Sustainable Development* 24(3): 190–198.

Baker, C. and Beaumont, J. (2011) *Postsecular Cities: Space, Theory and Practice*, London: Continuum.

Baker, C. and Dinham, A. (2017) 'New interdisciplinary spaces of religions and beliefs in contemporary thought and practice: an analysis', *Religions* 8(2): 1–16.

Bartkowski, J.P. and Swearingen, W.S. (1997) 'God meets Gaia in Austin, Texas: a case study of environmentalism as implicit religion', *Review of Religious Research* 38(4): 308–324.

Beebeejaun, Y. (2004) 'What's in a nation? Constructing ethnicity in the British planning system', *Planning Theory and Practice* 5(4): 437–451.

Beebeejaun, Y. (2012) 'Including the excluded? Changing the understandings of ethnicity in contemporary English planning', *Planning Theory and Practice* 13(4): 529–548.

Beebeejaun, Y. (2017) 'Gender, urban space, and the right to everyday life', *Journal of Urban Affairs* 39(3): 323–334.

Beebeejaun, Y. and Grimshaw, L. (2011) 'Is the 'New Deal for Communities' a new deal for equality? Getting women on board in neighbourhood governance', *Urban Studies* 48(10): 1997–2011.

Bowlby, S.R., McKie, L., Gregory, S. and MacPherson, I. (2010) *Interdependency and Care over the Lifecourse*, London: Routledge.

Braidotti, R. (2008) 'In spite of the times: the postsecular turn in feminism', *Theory, Culture and Society* 25(6): 1–24.

Cavanaugh, W.T. (1999) 'The world in a wafer: a geography of the Eucharist as resistance to globalization', *Modern Theology* 15: 181–196. doi: 10.1111/1468-0025.00093.

Dunlap, T.R. (2006) 'Environmentalism, a secular faith', *Environmental Values* 15(3): 321–330.

Gale, R. (2008) 'Locating religion in urban planning: beyond "race" and ethnicity?', *Planning Practice and Research* 23(1): 19–39.

Gale, R. and Naylor, S. (2002) 'Religion, planning and the city: the spatial politics of ethnic minority expression in British cities and towns', *Ethnicities* 2(3): 387–409.

Gilroy, R. and Booth, C. (1999) 'Building an infrastructure for everyday lives', *European Planning Studies* 17(3): 307–324.

Greed, C. (2011) 'A feminist critique of the postsecular city: god and gender' in C.R. Baker and J. Beaumont (eds) *Postsecular Cities: Space, Theory and Practice*, London: Continuum, 104–119.

Greed, C. (2016a) *Christianity and Planning: Pentecostal Churches*, Occasional Paper 1, University of the West of England, Bristol. http://eprints.uwe.ac.uk/29024.

Greed, C. (2016b) Religion and sustainable urban planning: 'If you can't count it, or won't count it, it doesn't count', *Sustainable Development* 24(3): 154–162.

Jupp, E. (2014) 'Women, communities, neighbourhoods: approaching gender and feminism within UK urban policy', *Antipode* 46(5): 1304–1322.

Jupp, E. (2017) 'Families, policy and place in times of austerity', *Area* 49(3): 266–272.

Locality (2018) *The Great British Sell Off: How We're Losing Our Vital Publically Owned Buildings and Spaces. Forever*, London: Locality: Save our Spaces.

Maddrell, A. (2011) 'Bereavement, belief and sense-making in the contemporary British landscape: three case studies', in C. Brace, A. Bailey, S. Carter, D. Harvey and N. Thomas (eds) *Emerging Geographies of Belief*, Newcastle upon Tyne: Cambridge Scholars, 216–238.

McClymont, K. (2014) 'Planning for the end? Cemeteries as planning's "skeleton in the closet"', *Town and Country Planning* 2014: 282–285.

McClymont, K. (2015) 'Postsecular planning? The idea of municipal spirituality', *Planning Theory and Practice* 16(4): 535–554.

McClymont, K. (2016) '"That eccentric use of land at the top of the hill": cemeteries and stories of the city', *Mortality* 21(4): 378–396. doi: 10.1080/13576275.2016.1151865.

McClymont, K. (2018) 'Spaces of secular faith? Shared assets and intangible values in diverse and changing communities', *Implicit Religion* 21(2). doi: 10.1558/imre.37013.

McDowell, L. (1983) 'Towards an understanding of the gender division of urban space', *Environment and Planning D: Society and Space* 1(1): 59–72.

McLennan, G. (2010) 'The postsecular turn', *Theory, Culture and Society* 27: 3–20.

McLennan, G. (2011) 'Postsecular cities and radical critique: a philosophical sea change?' in C.R. Baker and J. Beaumont (eds) *Postsecular Cities: Space, Theory and Practice*, London: Continuum, 15–32.

Mean, Melissa and Tims, Charlie (2005) 'People make places: growing the public life of cities', *Demos*, September. www.demos.co.uk/files/Demos_PMP_Final_02.pdf.

Metzger, J., Allmendinger, P. and Oosterlynck, S. (eds) (2015) *Planning against the Political: Democratic Deficits in European Territorial Governance*, London: Routledge.

Milbank, J. (2006). *Theology and Social Theory: Beyond Secular Reason*, Oxford: Wiley-Blackwell.

ONS (2016) 'Women shoulder the responsibility of "unpaid work"'. www.ons.gov.uk/peoplepopulationandcommunity/healthandsocialcare/healthandwellbeing/articles/fullstorythegendergapinunpaidcareprovisionisthereanimpactonhealthandeconomicposition/2013-05-16 [last accessed 13 December 2018].

Peterson, M. (2017) 'Living with difference in hyper-diverse areas: how important are encounters in semi-public spaces?', *Social and Cultural Geography* 18(8): 1067–1085.

Petersson, A. (2010) 'The production of a memorial place: materialising expressions of grief' in A. Maddrell and J. Sidaway (eds) *Deathscapes: Spaces for Death, Dying, Mourning and Remembrance*, Farnham: Ashgate, 141–160.

Primiano, L.N. (1995) 'Vernacular religion and the search for method in religious folklife', *Western Folklore* 54: 37–56.

Reilly, N. (2017) 'Recasting secular thinking for emancipatory feminist practice', *Social Compass* 64(4): 481–494.

Robinson, K. and Sheldon, R. (2018) 'Witnessing loss in the everyday: community buildings in austerity Britain', *The Sociological Review* 67(1): 111–125. doi: 10.1177/0038026118797828.

Sánchez de Madariaga, I. and Neuman, M. (2016) 'Mainstreaming gender in the city', *Town Planning Review* 87(5): 505–523.

Sandercock, L. (1998) *Towards Cosmopolis*, Chichester: John Wiley.

Sandercock, L. and Senbel, M. (2011) 'Spirituality, urban life and the urban professions' in C.R. Baker and J. Beaumont (eds), *Postsecular Cities: Space, Theory and Practice*, London: Continuum, 87–103.

Scott, J.W. (1996) *Only Paradoxes to Offer: French Feminists and the Rights of Man*, London and Cambridge, MA: Harvard University Press.

Scott, J. (2017) *Sex and Secularism*, Princeton, NJ: Princeton University Press.

Wilson, J. and Swyngedouw, E. (eds) (2015) *The Post Political and Its Discontents: Spaces of Depoliticisation, Spectres of Radical Politics*, Edinburgh: Edinburgh University Press.

Woodthorpe, K. (2011) 'Sustaining the contemporary cemetery: implementing policy alongside conflicting perspectives and purpose', *Mortality* 16: 259–276.

Worpole, K and Knox, K (2007) *The Social Value of Public Space*, York: Joseph Rowntree Foundation. www.jrf.org.uk/report/social-value-public-spaces [last accessed 3 September 2019].

11 Neo-paganisms, 'dark green religion' and what the divine feminine might mean for planning

Katie McClymont

Action as ritual: ritual as action.

(The Warrior's Call)

Abstract

In academic research increasing attention is being given to the areas of paganism, alternative spiritualties and 'lived religion' in the Global North, and how these issues and beliefs intersect with and shape notions of gender and feminism (Aune, 2015). However, the spatial implications, and questions of relationships with land and development remain substantively under-explored. The ways in which values and beliefs shape and are shaped by the natural environment, and understandings of the fragility of the earth offer different perspectives on religion, gender and 'the city'. This chapter introduces some of the main themes in relation to gender and planning from a perspective of paganism and 'spiritual but not religious' beliefs and practices. It outlines the importance and challenge of a Goddess-focused faith to thinking about gender and religion, and draws out how this brings different questions to the arena of urban planning. Specifically this covers notions of appropriate behaviour in pre-historic/sacred spaces, the value of nature and how this plays out in local politics and the role planning policy could have in engaging with (neo)Pagan spiritualties and practices.

Key words: Paganism, environment, values, Goddess, sacred space, planning

Introduction

Increasing attention has been given in academic research to the areas of Paganism, alternative spiritualities and 'lived religion' in the Global North (Harvey, 2016; Harvey and Hardman, 2000; Bowman, 2005; 2006; Sutcliffe, 2003; Sutcliffe and Gilhus, 2013; Davy, 2007) in the areas of sociology, geography and, unsurprisingly, religious studies. A resurgence in such spiritualities is often linked to a passion for environmental action, as exemplified in the Warrior's Call, quoted at the top of this chapter (see www.warriorscall.org). Further it is one of the fastest growing religious movements, with an estimation of between one and three million practitioners worldwide (Kraemer, 2012; Saunders, 2012: 787). As such, it can be seen as part of the diversity of the changing landscapes of contemporary faith; part of the 'post-secular turn' (McClennan, 2010). As a growing movement, it represents part of the rejection of the uniformity and uni-directionality of secularism, and attendant assumptions that religion would lose its important as a way of understanding and interaction in the world. Moreover, how these issues and beliefs intersect with and shape notions of gender and feminism has come under substantial debate (Aune, 2015; Kraemer, 2012; Mantin, 2000; Manning, 2012; Morgan, 2000). However, the spatial implications, and questions of relationships with land and development remain substantively under-explored (but see Saunders, 2012 on the religio-geography of Neopaganism). Unlike the majority of the other chapters in this volume, this piece does not draw a strong definition of religion or religious peoples to look at the intersections and problematics of planning, religion and gender in depth and detail. Instead, its main contribution is scoping a broad tranche of poorly defined practices, which are growing in importance because of both increasing numbers who fit this definition, and the sorts of conceptual and practical challenges they bring to the framing of usual categories deployed in understandings of religion and society. Because of this, the requirements or considerations for planning are different than for those of other faith groups. Instead of needing larger-scale premises, with adequate parking, which are well-located for worshipping communities to access, issues arise more about the value and subsequent treatment of nature and access to certain (pre)historical sites.

The chapter sets out a concise background to a broad range of practices and beliefs described in terms such as (neo)Pagan, nature-based religions, spiritual-but-not-religious (SBNR), New Age and indigenous (Harvey, 2016; Harvey and Hardman, 2000; Saunders, 2012). These terms are not without conflict and contestation – either by those in academia using them or those being described in such ways – but together they can be used to delineate a group of practices which are not secular, atheistic or profane, but do not conform to several of the usual expectations of 'world' religions.

Moreover, the specific ways in which values and beliefs of this group shape and are shaped by understandings of the natural environment are pertinent for this discussion. Taylor's (2010) notion of 'dark green religion' is of particular interest here. He describes it as 'like a phantom. It is unnamed and has no institutions officially devoted to its promotion' (2010: ix). Taking this alongside the centrality of the female divine/Goddess worship within much p/Pagan and other practices, exploring these practices offers ideas, insights and alternatives for issues relating to urban development, which are largely absent from understandings of other (predominantly monotheistic) world religions. Moreover, they offer wider challenges to the way religion is categorised and privatised by much of contemporary planning practice (see my other chapter, this volume).

For the purposes of this collection, this chapter aims to set out a background which will demonstrate the relevance of issues emerging from within these less prescribed spiritual movements – with a focus on what has been broadly defined as, and here is referred to as, (neo)Paganism and the questions of gender which are central within them – to planning and urban studies. This entire chapter could be filled with caveats of definition, apologies for imprecision, and semantic debate. I have chosen not to do this, as I do not think it is interesting or important enough to fill a chapter of this book on such matters. This therefore will leave my writing open to accusations of imprecision, oversimplification and conflation – which will at several points be no doubt justified. To try to slightly mitigate against this, without distracting from the flow of the overall chapter, I will now give a brief explanation for my choice of terms. I have put 'neo' in parenthesis to indicate that some people would not be happy with this prefix; however, others would not accept the term Paganism to describe their beliefs, but wish to use this terms as a general shorthand to describe a range of beliefs all of which have fluid definitions. I have chosen to use Paganism with a capital P as 'with a lowercase p, "paganism" usually refers to all religions that have resisted conservation to monotheistic traditions' (Davy, 2007: 2) whereas Paganism with a capital refers to the modern movements. When it is grammatically possibly, I use the term (neo)Paganisms as this reflects the diversity of practices more accurately. Further, to reflect on my own positionality within this, as is fitting in a feminist-inspired volume, I have a personal interest in such practices, and these ways of viewing both nature and the divine feminine, but like many, find these definitional terms insufficient or inaccurate in describing my beliefs. The chapter aims to develop a research agenda, showing how such concerns are of importance in themselves to planning practice, but also in dialogue with other religions, and with a broader sense of spirituality or 're-enchantment' in society today. It does so by highlighting three different case studies of planning issues and their (lack of) engagement with substantive (neo) Pagan concerns and identities. It then reflects on these through a feminist lens, and one of men and women's spiritualties, and their role in modern development. The aim here is not to say anything conclusive. It is to open up new ways of thinking about what counts as a religion in the fields of planning and urban studies, and to demonstrate why a consideration of (neo)Paganism is important in this area, and important for considerations of the complex tryptic of gender, religion and the urban.

Mapping the landscape SBNR/paganisms

There is no one overarching theology or belief system which unites groups or individuals who are classified as anything from spiritual seekers to members of more established Pagan groups such as Druids; it is 'as much defined by its diversity, heterogeneity, and lack of structure as its commonalities' (Saunders, 2012: 789). However, there are some core tenets which can be ascribed both more specifically to (neo)Paganism, and to SBNR practices more generally. To understand this, there are a number of key issues which need raising here which shape both how such identities and movements intersect with ideas of gender, the environment and with planning. The section will briefly consider issues about the debates over the difference between 'religions' and 'spirituality' as well as defining more fully some of the history and features of these groups and movements.

Most academic histories of contemporary (neo)Paganism date it to the mid-nineteenth century, with a few earlier influences, especially the establishment of Druidic Orders in the 1700s, being formative in its development (Hutton, 2000; Rountree, 2012). However, many practitioners claim a much longer heritage, stating that their practices and beliefs are part of a lost 'indigenous'/pre-Christian religion, part of long supressed 'old ways', linking in particular to notions of 'wise women' or witches and witchcraft, as well as invocations of Celtic goddesses. Most academic authors assert that there is little evidence for this, but the influence and ongoing importance of 'prehistoric' sites is an important part of modern (neo)Paganism (Blain and Wallis, 2004; Cusack, 2012), as discussed at greater length below. What is clear is that what is now classified as modern (neo)Paganism, or 'New Age' spirituality springs from a wide range of eclectic roots and traditions: 'modern Paganism was neither the descendant of a continuous sectarian witch cult, nor born fully-fledged from the imagination of one man (Gerald Gardener) in the 1940s' (Hutton, 2000: 13).

With this background in mind, it is evident that simple, well-defined categories will be absent from the understanding of (neo)Paganisms discussed here. Moreover, considering the myriad sets of beliefs and practices which fit into this category raises important questions about the definition of 'religion' (York, 2000). Because practices, theologies and structures are so different from Judeo-Christian-Islamic ones, and bear only very passing associations with Hindu or Buddhist theologies, we '(require) a new optic to understand how place and space are contextualised within various systems of belief' (Saunders, 2012: 789) within (neo)Pagan and related spiritualties. Saunders (2012) further notes the central importance of place in (neo)Pagan beliefs and practices; because of the lack of a core text, or written set of beliefs, sacred sites hold a more prominent role in issues of identity and faith than in some other religions. This in turn means that planning issues – those related to the management, maintenance and designation of places – will intersect with (neo)Pagan practices in ways which differ markedly from seeking permissions for places of worship. Because these broader spiritual beliefs challenge pre-set categories of 'religious' and 'secular', they bring new and different issues to thinking about planning practices which are discussed at greater length below in relation to ideas of gender and of the environment. Before doing this, however, it is necessary to consider more fully the implications of this blurring of definitions of religion.

Much of the literature argues that the increasing modern interest in 'spirituality' in the face of declining attendance and membership of organised religions parallels an increasing privatisation and individualisation as a trend in modernity, expressing that 'religion and spirituality maps relatively easily onto [the distinction] between public and private' (Harvey, 2016: 130). However, Harvey (2016) also goes on to argue that this is simplistic, and does not reflect the range of practices and activities which can be classified under the heading of 'spirituality'. Although there is no one set of theological cannons which co-ordinate the beliefs of any of the (neo)Pagan practices which this chapter considers, there are central concerns (especially concerning gender and the environment) which bring this grouping together: '(i)n a wide range of indigenous and other communities, key questions are asked about what constitutes good ways to live alongside other-than-human persons' (Harvey, 2016: 134). This means that a simple division between 'private(-ised) spiritualties' and 'public/communal religion' is not entirely accurate, and obfuscates collective concerns which can be seen as emerging from these practices. These include challenges stemming from the ideas of immanent Goddess or divine feminine worship and the changed way this leads practitioners to view the world. Unsettling Durkheim's claim that religion's role is in maintaining society and its moral cohesion, Rountree argues that 'modern Paganism can be seen as a protest against, rather than a projection of, society' (Rountree, 2012: 306). In turn, it could be argued that (neo)Paganism is a project of what society *should* be – with more equitable gender relations and attendant reverence for nature and the environment. This is discussed in greater depth below.

Gender and the Goddess

Goddess worship, and the idea of feminine divinity is central to many if not most branches of (neo)Paganism, especially Wicca. As with the majority of (neo)Paganisms, there are very different, personal and subjective ways that this has been interpreted but it remains a powerful aspect of both individual spiritual practices and the wider gendered outlook of many (neo)Pagans. In academic writings, the influence of Carol Christ's (1979; 2011) work, especially her notion of *thealogy*, and her anti-monotheistic feminist writings (Mantin, 2000), is notable. These explore the importance of reframing theologies from a feminist perspective, outlining the 'regenerative power of reclaimed language and symbolism' (Rountree, 2012: 148) in (re)establishing ideas of female deity. This reframing challenges the divide between monotheism and pantheism – allowing both for 'the Goddess' and multiple goddesses, and opens (neo)Pagan worship up to a plethora of traditions on the grounds of claiming a holistic understanding of the divine (Morgan, 2000; Mantin 2000). This in itself is seen as important by practitioners and writers: '(t)he Goddess at a subliminal level of our awareness is a very powerful force for motivation and change' (Morgan, 2000: 94). The very notion of a female divine presence, however subjectively interpreted and therefore turned into personal practice, is fundamental in the way (neo)Paganism and many related SBNR beliefs operate and therefore encounter issues of cities and planning. Not only does it raise the importance of women as practitioners, and therefore actors in the public realm, it raises the importance of female bodily experience: childbirth, lactation, menstruation are all part of the cyclical nature of womanhood (and therefore in the image of the Goddess as maiden/mother/crone), and therefore fundamental, and fundamentally positive, parts of life and to be celebrated rather than hidden away, suggesting a radical new (feminist) world order.

Moreover, Goddess worship not only rebalances or transgresses established gendered hierarchies, it also brings in a different way of understanding the world, something which is followed through in (neo)Pagan theologies of nature and the environment: 'Goddess, as the sacred *immanent* in the natural world, expresses the teeming biodiversity of all forms of life and death' (Mantin, 2000: 148, emphasis added). This is expressed through (neo)Pagan descriptions of the natural world, with commentators 'tending to feminise the landscape, seeing rounded hills as the earth Goddess's breasts and caves as Her womb' (Rountree, 2012: 308). This sense of the immanence of the divine is central to both ideas of the environment and ideas of gender, at least partially because of the lack of division in (neo)Pagan ideas between humans and the rest of the 'non-human' natural world. It is counter to a notion of an entirely transcendent deity who cannot be apprehended directly in this world, and therefore whose sacredness is not part of nature or embodied human existence. There are parallels to be drawn here with critiques of rationalist planning (Sandercock, 1998; 2003) and to ecofeminist movements and value systems (Mann, 2011): (neo)Pagan voices resonate with the embodied, emotionally sensitive approach to cities promoted by authors such as Sandercock, but take this to include all of the 'feeling' world, not just that of humans, but also seeing the power dimensions within environmental exploitation.

The issues here are therefore different to those around gender and religion in many other chapters discussed in this volume. Because (neo)Paganism can be viewed as majority female, with a strong sense of divinity as either female or encompassing both/many genders (Kraemer, 2012), its practice for women, and theology, is more directly linked, rather than seeing/searching for female space within a patriarchal practice. This is reflected in the (neo)Pagan approach to nature.

The environment/nature and the spatial as sacred

As fundamental to (neo)Pagans as the divine feminine is the notion that the earth is sacred.

> The notion of society as a network of *conscious, agentic, related, mutually dependent and responsible beings, of who only some are human*, represents a projection of Pagan philosophical and political values and claims regarding human–environmental relationships, including ideas about how power, agency and responsibility should be negotiated and enacted within these relationships.
>
> (Rountree, 2012: 306, emphasis added)

This is linked to a rejection of both Christian privileging of humans over other living beings and of European philosophy's premise that 'whatever cannot be said simply does not exist' (Harris, 2000: 152). This is clearly a vast sweeping generalisation of both Christianity and Western philosophy but does demonstrate how (neo)Pagan notions of nature – situated within countries with dominantly Judeo-Christian and/or liberal secular traditions and state structures – differ markedly from their dominant cultures. Again, practices vary greatly over issues such as eating animals (Harvey, 2016; Rountree, 2012) and how this can/cannot be justified, but the overarching philosophical premise here is markedly different to standard Western ontologies. This challenges established ways of knowing and being and, in turn, the role this new ontology brings to action: 'what is required is another way of knowing, a Sacred Ecology which moves beyond the cerebral to bring us to a direct experience of a wholeness rooted in the body' (Harris, 2000: 151). This is not only a theological or ontological issue: it is something which can have a marked impact on political motivations and activism, summed up well by writer and founder of Dragon Environmental (Dragon Environmental Network, 2015), Adrian Harris: 'We act to protect our Earth because we know, in every cell of our bodies, that our lives, our communities and our land are scared' (Harris, 2000: 153). In the same way that notions of the female divine challenge and inspire a different understanding of divinity among and within women, this too can be a deeply subjective, lived set of beliefs which not only challenge ways of thinking and doing, but also challenge the relationship between the two.

What are the questions for planners here?

This section will now discuss three different examples to begin to suggest some of the ways in which the concerns raised above about gender and (neo) Paganisms resonate more specifically with contemporary planning issues. Again, it is important to note that these are presented to suggest avenues of interests for further exploration, rather than any substantive research findings. The first discusses the contested nature of 'sacred sites' or 'prehistoric heritage' and how practices of both their management and the attendant ways in which they are valued are challenged by acknowledgement of the credibility of (neo)Paganisms as genuine religious beliefs. The second covers issues of environmental activism, and the implications of viewing the earth/nature as sacred in local political issues, as well as in lifestyle choices. The third looks at the local planning policies for one of the central locations of contemporary (neo)Paganisms – Glastonbury in Somerset (cf. Bowman, 2005; 2006) – to demonstrate both the current lack of engagement with this aspect of the place, and to suggest the sort of things which could be done differently if they were to be.

Sacred sites?

The contestation of the value and attendant legitimate practices in, on and around 'sacred' or 'prehistoric' sites forms one of the most well-documented encounters of (neo)Paganisms with planning practice. Although largely in rural rather than urban contexts, it is worth briefly summarising this debate here as it demonstrates several important issues about relevant issues for this volume. The importance of religious values in the management of 'natural' settings is debated by Kraft (2010) in relation to Tromsdalstind in Norway. It is a mountain which was proposed as a site for downhill skiing in the Winter Olympics; however, this proposal was rejected after it was declared a Sami sacred place. Kraft argues that 'sacred places provide an alternative mapping – an appropriation of a particular landscape ... (and that) such demarcation(s) may ... have practical consequences' (Kraft, 2010: 57). This discussion frames the debates of how (neo) Pagan spiritualties and practices have clashed with official interpretations of the meanings, values and therefore legitimate practices. In the UK, sites including Stonehenge, Avebury and the Rollright Stones have demonstrated different levels of tolerance and conflict over their ability to hold a dissonant interpretation of their value; as Cusack states 'The interpretations of Stonehenge have been difficult, if not impossible, to police' (Cusack, 2012: 152). Many (neo) Pagans in the UK see Stonehenge as a sacred site of central importance to their beliefs, and as a link to lost Pagan practices of a pre-Christian era. Because of this, their interpretation of the site is as an active place of worship – somewhere to be present at during key festivals (the Summer Solstice/21 June being the most important to most practitioners) and to invoke religious rites differs markedly from the official interpretation (and attendant legitimate practices) of its formal custodians: Historic England. There is a long history of battles for access from the 1960s onwards, interweaving with New Traveller, 'rave' culture and fears of civil disobedience (McKay, 1998). As Saunders states, 'For Pagans, interacting with the sites often involves drinking and dancing *in* and often *on* places an archaeologist would consider off-limits' (Saunders, 2012: 798). This clash of interpretation over appropriate behaviour belies a deeper clash of epistemologies:

> One effect (of seeing 'spirits within the stones' as irrational) is to further marginalize and trivialize 'spirit' discourses, which, together with interpretations based on them, are invisible within formal interpretations or public representations of landscape.
>
> (Blain and Wallis, 2004: 243)

Blain and Wallis' (2004) discussion of the clashes between (neo)Pagans and official Historic England understandings of Stonehenge (and other related sites) demonstrates the marginalisation of (neo)Pagan beliefs that the earth is sacred, and that this sacredness is immanent – and female. This demonstrates how rationalist, 'scientific' stances over the management of the built and natural environment disallow certain (predominantly female) spiritual interpretations of places. Moreover, this is not just a matter of knowledge, but of the sorts of worship which are therein permitted. 'In the case of sites of religious or spiritual significance, this stance frequently results in clashes between the "scientific" interpretations proposed by archaeologists, and theological interpretations of archaeological material developed by faith communities' (Cusack, 2012: 145).

Blain and Wallis go on to discuss the Rollright Stones, a less-well known prehistoric stone circle in Oxfordshire, England, wherein ownership and management has been transferred to a partially Pagan organisation, creating guardians who can transcend the heritage/spirituality chasm 'through the direct and intensive work of a number of people who have set out to accomplish shared meaning' (Blain and Wallis, 2004: 256). This demonstrates that there can be ways around clashing interpretations which promote practices which please both archaeological and (neo)Pagan interests. It would be interesting to explore directly the role gender plays in this, both in terms of the practitioners and activists involved in the running and custodianship of the site, as well as in the minds of visitors, but also in the way(s) in which shared meanings were established, and whether wider lessons about the management of prehistoric sacred sites be transferred more widely.

Environmental activism?

As is evident from the preceding discussion, issues of the environment are key to contemporary (neo)Paganisms. Harris (2000) cites the Dragon Environmental campaign group and its historical involvement in activism against road building and other desecrations of sacred places (see also Dragon Environmental Network, 2015). In terms of gender, this resonates both theologically and practically. For many practitioners, the earth is sacred as she is 'Mother' – the immanence of the divine is within all things and at least equally female as male. Moreover, many of these practitioners and activists are female, as is demonstrated with the two examples below.

Fracking remains one of the most contentious environmental practices in both the UK and the US, its science, impact and desirability both locally and globally being highly contested by different groups (Beebeejaun, 2017). It is therefore quite expected that modern (neo)Pagans have responded both actively and negatively to this practice. This can be seen in *The Warrior's Call: Pagans United against Fracking*, whose promoters claim 'Earth, Sea, and Sky are threatened by the commitment to unending economic growth powered by fossil fuels without regard for the consequences' (The Warrior's Call, 2019). The maxim included at the start of this chapter 'Action as Ritual: Ritual as Action' is the Warrior's Call, which epitomises the proactive nature of the movement (to put it in Christian terms, not just faith but works too) This campaign is not only home to updates about anti-fracking protests, useful guides to the process of fracking and a blog by activists but also ways in which spiritual practices and rituals can be harnessed to challenge the practices of fracking, such as 'Waters of the World: place blessings and healing in the water courses of your local area. Let them flow into the oceans to protect the waters of the world' (The Warrior's Call, 2019). The phrase 'Action as Ritual: Ritual as Action' drawn upon in the title of this paper comes from them; demonstrating how politically motivated deeds are concomitantly religious ones, and that the purpose of faith-inspired practices are to bring about positive change in the world. A contention that the earth is sacred logically leads to rejection of and active resistance to practices which seek to harm this sacred being, questioning and reframing claims that such practices could be in the public interest. More broadly, anti-fracking movements demonstrate 'a dissatisfaction with the limitations of current regulatory processes that are held within the confines of traditional notions of expertise and not open to wider public engagement' (Beebeejaun, 2017: 428). This demonstrates both the current failure of planning policy and practices to be open to wider epistemologies, but also the possibilities that such new frameworks could bring by envisioning other ways of valuing the earth. In this way, (neo)Paganisms deny the possibility and desirability of seeing this in purely human terms, harking to a more holistic ontology which shares concerns with that of ecofeminism and its engagements with spirituality and postsecularism. (Braidotti, 2008; my other chapter, this volume).

This same reasoning can be seen in a more localised and small-scale example. In Sheffield, Yorkshire, since 2014 (STAG, 2019) there has been much controversy over the felling of several street trees as part of a privatised maintenance contract the city council had with contractors Amey. Unsurprisingly this has provoked widespread resistance and the campaign against this has drawn together a wide range of support, including the local (neo)Pagan community. This is seen in the ongoing connections through Airy Fairy, a Pagan shop and café which also forms a focal point for local and regional Pagan events (see The Tree Council, 2019 for example). In this context, shops/cafés do not simply play the role as retail/food outlets – they have a wider community centre/place of worship focus for a faith/collection of practices which do not have formal spaces for worship in most settings. Airy Fairy has also played a wider role in local politics, using its prominent shop window to campaign for Green Party candidates at local election time (Teal, 2018). This demonstrates how religious faith can intersect with political activism, a faith and an activism which both have strong pro-feminist and women-centred ideologies as well as membership. Alongside the previous discussion about fracking, these demonstrate that (neo)Pagan practices have a role in contemporary planning debates, one which can bring new and radical challenges to the way debates about the environment are framed. This spans daily city life, seen in the value attached to street trees and urban wildlife, to off-grid/low-impact living (Fairlie, 2009). The motivations of some who choose to live beyond the usual system, whether in 'intentional communities' (Sager, 2018) or alone/in family smallholdings at times relates to claims to credibility which are made on value systems outside that of mainstream society (McClymont and Sheppard, 2019). These claims often revolve around an idea of the importance, and at times sacredness of nature and the earth. This is also an area which needs to be researched at greater depth to understand the dynamics and issues at stake here.

Planning policy

The final example given here is the most speculative, but aims to demonstrate how (neo)Pagan concerns, and their attendant gendered aspects, relate to everyday planning practice. Below is an extract from Mendip council's local plan (Mendip District Council, 2014) about the role of Glastonbury town centre in its development strategies. As is well documented elsewhere, 'Glastonbury's current status as a significant and sacred site for a great range of people is unparalleled in the British Isles' (Bowman, 2006: 124). As well as being home to the historic Glastonbury Abbey and Glastonbury Tor, it hosts several Pagan/Occult/New Age shops, Chalice Well – 'a place where people have gathered to drink the waters and find solace, peace and inspiration' (Chalice Well Trust, 2019) for over 2000 years, and the Goddess temple (Glastonbury Goddess Temple, 2019), bringing many visitors both locally and internationally. However, this is not evident in local policy:

> 5.14. It (Glastonbury Town Centre) is dominated by independent shops and traders and its alternative shops are a feature of the town reflecting cultural and spiritual aspects which appeal to tourists and visitors. Residents feel that the independent shops are important to the town and reflect its personality although the latest town centres study suggests that their increasing numbers may be squeezing out more typical retail activity, reducing the overall amenity of the centre to local residents.
>
> (Mendip District Council, 2014: 58)

Council planning policy seems at best a little embarrassed by the town's status, and does not reflect its value in any way other than economically; and this is only seen as partially positive. It is not the case that planning policy cannot demonstrate any appreciation of the religious or spiritual value of places. North Norfolk council's conservation area appraisal for Walsingham directly states that 'Walsingham is best known for its historic and present day significance as a pilgrimage centre' (North Norfolk District Council, 2013: 4) going on to detail 'the *vision* that transformed Walsingham' (North Norfolk District Council, 2013: 6, emphasis added) discussing the story of Richeldis de Faverches' visions of the Virgin Mary, and how this shaped the place ever since. It goes on to detail how this shaped its historical development, the establishment of a Franciscan friary, then hotels and other services for pilgrims. Although neither the places nor the policies are directly comparable, this does highlight the invisibility of (neo) Pagan spiritualties as something of worth, value or even note to planning practitioners. Planning for the town of Glastonbury seems like an excellent opportunity to attempt to see what policy supportive of Goddess worship could look like in practice. Again, this area warrants further research and consideration.

Conclusions/other thoughts

This chapter has begun to outline some of the contours of the landscape of (neo)Paganisms, including SBNR beliefs in terms of gender and planning. It outlines that there is very limited engagement in this area, but the topic raises several important considerations for planning – more widely than just about the importance of taking seriously the needs and concerns of a range of eclectic practices and practitioners. Questions about the value of nature and of the earth, stemming from a belief in an immanent feminine divine, challenge many set assumptions based on notions of the earth as a resource for humans, underpinned by ideas of ecological modernisation (Dobson, 2006; Taylor, 2010) as hinted at in the discussion of (neo)Pagan environmental activism. The importance of prehistoric practices of going to ancient sites 'is one way in which neo-Pagans address the dislocation of alienation from nature and ancestral kin they experience as members of contemporary Western societies' (Rountree, 2012: 315).

The importance of these challenges to interpretations of environmental politics and the management and value of prehistoric sacred sites rests on questions of the value, authenticity and credibility of (neo)Paganisms as a religion. If all religion is 'licenced insanity' (Bowker, 1987, cited in Harvey, 2016), why should some matter more than others? There is an important debate to be had about cultural appropriation and about authenticity of (any/religious) practices. This is not the place to do so, but I feel it is important to acknowledge these issues rather than appear to ignore, occlude or dismiss them. There are good grounds to question the appropriateness of some practices which make universalising assumptions about 'native peoples', and their inherent value as somehow closer to nature and therefore more spiritually attuned (Mann, 2011: 18–19). Even if (neo)Pagan beliefs can be classed as elective, rather than culturally inherited (Rountree, 2012), it is a non sequitur that this means they are inauthentic or not worthy of being taken seriously as religious practices. Moreover, such a dismissal could be seen as a perpetuation of patriarchal denial of female notions of the divine and the environmental implications of an immanent religion. This is summed up well by Rountree (2012):

> A desire to re-enchant the earth and invest it with sacred significance framed in anthropomorphic and anthropocentric terms may indeed be a religiously appealing, politically motivated neo-Romantic critique of modernity, 'progress' and capitalism, a response to increasingly secularised, technology driven and rather mundane lives. Even so, many Pagans would argue, an ontological reality of interconnectedness, endorsed by quantum physics, underlies their worldview.
>
> (Rountree, 2012: 316)

This position is a profound challenge to many of the norms of planning practice, as well as contemporary gender relations. By accepting (neo)Pagan views as legitimate, planning can begin to engage differently with issues of place management, prehistory and the environment.

References

Aune, K. (2015) 'Feminist spirituality as lived religion: how UK feminists forge religio-spiritual lives', *Gender and Society* 29(1): 122–145 doi: 10.1177/0891243214545681.

Beebeejaun, Y. (2017) 'Exploring the intersections between local knowledge and environmental regulation: a study of shale gas extraction in Texas and Lancashire', *Environment and Planning C: Politics and Space* 35(3): 417–433.

Blain, J. and Wallis, R. (2004) 'Sacred sites, contested rites/rights: contemporary pagan engagements with the past', *Journal of Material Culture* 9(3): 237–261.

Bowman, M. (2005) 'Ancient Avalon, New Jerusalem, heart chakra of planet earth: the local and the global in Glastonbury', *Numen* 52(2): 157–190.

Bowman, M. (2006) 'The Holy Thorn Ceremony: revival, rivalry and civil religion in Glastonbury: presidential address given to the Folklore Society, March 2005', *Folklore* 117(2): 123–140.

Braidotti, R. (2008) 'In spite of the times: the postsecular turn in feminism', *Theory, Culture and Society* 25(6): 1–24.

Chalice Well Trust (2019) www.chalicewell.org.uk/ [last accessed 26 March 2019].

Christ, C. (1979) 'Why women need the Goddess: phenomenological, psychological and political reflections' in C. Christ and J. Plaskow (eds) *Womanspirit Rising: A Feminist Reader in Religion*, San Francisco, CA: Harper and Row, 273–287.

Christ, C.P. (2011) 'The last dualism: life and death in goddess feminist thealogy', *Journal of Feminist Studies in Religion* 27(1): 129–145.

Cusack, C. (2012) 'Charmed circle: Stonehenge, contemporary Paganism, and alternative archaeology', *Numen* 59(2/3): 138–155.

Davy, B.J. (2007) *Introduction to Pagan Studies*, London: AltaMira Press.

Dobson, A. (2006) *Green Political Thought*, London: Routledge (4th edition).

Dragon Environmental Network (2015) www.dragonnetwork.org/ [last accessed 26 March 2019].

Fairlie, S. (2009) *Low Impact Development*, Charlbury: John Carpenter Publishing (2nd edition).

Glastonbury Goddess Temple (2019) https://goddesstemple.co.uk/ [last accessed 26 March 2019].

Harris, A. (2000) 'Sacred ecology' in G. Harvey and C. Hardman (eds) *Pagan Pathways: A Guide to the Ancient Earth Traditions*, London: Thorsons, 149–156.

Harvey, G. (2016) 'If "spiritual but not religious" people are not religious what difference do they make?', *Journal for the Study of Spirituality* 6(2): 128–141.

Harvey, G. and Hardman, C. (2000) *Pagan Pathways: A Guide to the Ancient Earth Traditions*, London: Thorsons.

Hutton, W. (2000) 'The roots of modern paganism' in G. Harvey and C. Hardman (eds) *Pagan Pathways: A Guide to the Ancient Earth Traditions*, London: Thorsons, 3–15.

Kraemer, C.H. (2012) 'Gender and sexuality in contemporary Paganism', *Religion Compass* 6: 390–401. doi: 10.1111/j.1749-8171.2012.00367.x.

Kraft, S. (2010) 'The making of a sacred mountain: meanings of nature and sacredness in Sápmi and northern Norway', *Religion* 40(1): 53–61, doi: 10.1016/j.religion.2009.08.011.

Mann, S.A. (2011) 'Pioneers of U.S. ecofeminism and environmental justice', *Feminist Formations* 23(2): 1–25.

Manning, L. (2012) 'Experiences from Pagan women: a closer look at croning rituals', *Journal of Aging Studies* 26: 102–108.

Mantin, R. (2000) 'Thealogies in process: re-searching and theorising spiritualities, subjectivities and Goddess-talk' in J. Blain, D. Ezzy and G. Harvey (eds) *Researching Paganisms*, Walnut Creek, CA: AltaMira Press, 147–170.

McClymont, K. and Sheppard, A. (2019) 'Credibility without legitimacy? Informal development in a highly regulated context', *Cities* (forthcoming).

McKay, G. (1998) *DIY Culture: Party and Protest in Nineties Britain*, London: Verso.

McLennan, G. (2010) 'The postsecular turn', *Theory, Culture and Society* 27: 3–20.

Mendip District Council (2014) *Mendip District Local Plan 2006–2029*. www.mendip.gov.uk/CHttpHandler.ashx?id=9073&p=0 [last accessed 26 March 2019].

Morgan, L. (2000) 'Women and the Goddess to day' in G. Harvey and C. Hardman (eds) *Pagan Pathways: A Guide to the Ancient Earth Traditions*, London: Thorsons, 94–108.

North Norfolk District Council (2013) *Walsingham Conservation Area: Character Appraisal and Management Proposals*. www.north-norfolk.gov.uk/media/1458/walsingham_ca.pdf [last accessed 26 March 2019].

Rountree, K. (2012) 'Neo-Paganism, animism, and kinship with nature', *Journal of Contemporary Religion* 27(2): 305–320. doi: 10.1080/13537903.2012.675746.

Sager, T. (2018) 'Planning by intentional communities: an understudied form of activist planning', *Planning Theory* 17(4): 449–471.

Sandercock, L. (1998) *Towards Cosmopolis: Planning for Multicultural Cities*, Chichester: Wiley.

Sandercock, L. (2003) *Cosmopolis II: Mongrel Cities in the 21st Century*, London and New York: Continuum.

Saunders, R. (2012) 'Pagan places: towards a religiogeography of Neopaganism', *Progress in Human Geography* 37(6): 786–810.

STAG (2019) Sheffield Trees Action Group. https://savesheffieldtrees.org.uk/ [last accessed 26 March 2019].

Sutcliffe, S. (2003) *Children of the New Age: A History of Spiritual Practices*, London: Routledge.

Sutcliffe, S. and Gilhus, I. (eds) (2013) *New Age Spirituality: Rethinking Religion*, Durham: Acumen.

Taylor, B. (2010) *Dark Green Religion: Nature, Spirituality and the Planetary Future*, Berkeley and Los Angeles, CA: University of California Press.

Teal, A. (2018) https://twitter.com/alisonclareteal/status/991678978825117698.

The Tree Council (2019) 'Tree Huggers: an exhibition of illustrations inspired by Sheffield street trees'. www.treecouncil.org.uk/Take-Part/Event-Details/EventId/1458 [last accessed 26 March 2019].

The Warrior's Call (2019) www.warriorscall.org/ [last accessed 26 March 2019].

York, M. (2000) 'New Age and Paganism' in G. Harvey and C. Hardman (eds) *Pagan Pathways: A Guide to the Ancient Earth Traditions*, London: Thorsons, 157–165.

Part V

Planning policy and change

12 The planning system and places of worship in relation to the Anglican Church

Andrea Mulkeen

Abstract

This chapter examines how the planning system in England after the Second World War recognised and addressed the need for space for places of worship, from the planning of New Towns, expanded towns and overspill areas, to more recent major residential developments. The chapter comprises an historical account, based on research completed in 2014, which gives a detailed account of how religious issues were increasingly marginalised in the late twentieth and early twenty-first century. While recognising the significant changes in the English religious landscape since 1945, the primary focus is the extent to which the planning system has enabled the Church of England, as the Established Church, to maintain a presence in every community.

Key words: Anglican churches, planning law, post-war reconstruction, new towns, neighbourhoods

Editor's explanation

This chapter is based upon research undertaken 2010–2014 (Mulkeen, 2014), illustrated by case studies, which charts how religious and church matters were increasingly marginalised by the planners, in the name of modernisation and streamlining of the planning system. Andrea's meticulous research revealed the way in which this was happening. Therefore no attempt is made to update the references in this chapter, as it stands in its own right, but other chapters provide further insights on 'what happened next' (particularly the chapter by Carey Doyle) and direction to more recent literature on 'the same old problems' is to be found within other chapters. For example the planning consultancy CAG is still going strong (CAG, 2017), a critique of the planning system vis-à-vis its neglect of the spatial needs of faith groups continues (Rogers and Gale, 2015), particularly in respect of Pentecostal churches (Greed, 2016), while the nation as a whole continues to be less formally religiously observant but still values personal spirituality (Davie, 2015).

We now return to Andrea's chapter written in her own words.

Introduction

Successive governments advocate an important role for faith groups in civil society, acknowledging their significant contribution towards community building and voluntary and charitable sectors. Does the planning system facilitate this role? This chapter provides a historical perspective on the extent to which, since the Second World War, the planning system has recognised and addressed the need for worship space. It focuses on the Church of England's experiences, particularly in areas of significant new development and population growth.

Literature review

Much academic literature focuses, unsurprisingly, on issues facing growing minority faith groups navigating the planning system to meet their space needs, underlining increasing diversity in the contemporary English religious landscape. Peach and Gale (2003) describe challenges to the planning profession reconciling conservation of the existing iconography of urban landscapes with attempts by different groups to signify their identity through their faith buildings. Recognising the Church of England's unique legal status as the Established Church, Eade (2011) concludes that it retains a dominant position symbolically in the public imagination, in relation to which other religious groups are assessed.

Evaluations of the struggles of growing Pentecostal and charismatic congregations to find churches, particularly in London (Burgess, 2009; Garbin, 2013), identify a need for spacious, flexible premises accommodating a wide range of educational and welfare activities additional to worship use. Greed (2011) notes that attempts to achieve a change of use for large, empty industrial units on city edges have been thwarted by planners citing potential noise, disturbance or car parking requirements, even when no-one is living nearby, concluding that without a specific 'worship' land-use category, they are judged under laws created to deal with places of assembly, cinemas and dance halls and found wanting. Outward political interest in faith communities, within diversity and equality agendas, is contrasted with an apparent lack of responsiveness, in urban decision making and regeneration, to the spatial needs of many non-traditional congregations. Rogers (2013) criticises the paucity of references to faith groups in planning policy, with those there are failing to capture specific needs and contributions. Overall, the role of the planning system in relation to social identities, including religion, not easily conflated with 'ethnicity' or 'race', is insufficiently explored (Gale, 2008).

The Church of England's experiences and building needs are perhaps atypical. An inherited responsibility for 16,000 mainly historic churches stimulates keen interest in conservation and heritage control (Mynors, 1985; Mansfield, 2007) or dealing with over-provision (Comedia, 1995) rather than focusing on contemporary needs for new worship space. Analysis of such provision is primarily found in accounts of church building in post-war New Towns (Osborn and

Whittick, 1977; Baker, 2002), highlighting the role ascribed to churches in fostering new communities, notably where early emphasis on housebuilding meant that community facilities fell behind population growth (Ward, 1993). The Church's experiences of other major housing growth through urban extensions and new settlements are less well documented, even though mass migration was identified as the greatest social fact challenging the Church's parochial system after 1945 (Chandler, 2006).

Separately, notwithstanding decline in church membership and attendance of major Christian denominations, there is significant political discourse on the role and influence of faith groups (Brierley, 2016) informing debates on public policy initiatives, including community cohesion and urban regeneration, despite reservations about promoting the idea of an objective, homogenised faith sector. According to Chapman and Lowndes (2008), the 'faith sector' is constituted through the new governance strategies of urban policy makers, but difficulties of generalising within and across different groups raise challenges for engagement with planning structures. Deakin (2010) notes that religious groups allege lack of understanding or prejudice on the part of secular agencies, while officials express anxiety about proselytising religions and separatism, with concerns about possible sectarianism and exclusivity informing debate about the appropriate role of faith in the public realm. The growth in faith-based organisations providing social services (Ghatak and Abel, 2013) encourages an interest in community roles than needs of worshipping communities.

Farnell et al. (2003) highlight significant inequalities between faith groups in their abilities to engage in and resource community work. While some can offer large, adaptable spaces for wider benefit, adjacent or linked to places of worship, many operate in old, ill-adapted buildings. The ubiquity and established status of the Church of England, its networks, resources, parochial organisation, diocesan structure and hierarchy is seen as instrumental in the prominence of Anglican clergy and lay members within formal regeneration partnerships.

Post-war planning and meeting the need for new places of worship

In the immediate aftermath of the Second World War, specific provision was made for churches and other places of worship in the planning process for New Towns, less so in other areas with significant new housing.

War time and reconstruction

As the Established Church, the Church of England seeks to maintain a presence in every community. A wartime Church Commission, echoing widespread dissatisfaction with the limitations of inter-war planning, highlighted 'the vast extension of those spiritual and cultural wildernesses which have spread like a "red belt" round so many of our cities and towns' (Church Assembly, 1944: 4).

Lack of community facilities on large municipal estates was criticised, with the cost of providing churches on poorly positioned sites deemed prohibitive. Envisioning self-sufficient neighbourhoods mirrored utopian secular planning opinion, the belief in an active, interventionist role in planning new communities (Cullingworth and Nadin, 2006).

Lord Reith, Chairman of the New Towns Committee, corresponding with the Archbishop of Canterbury, saw an opportunity to foster moral and social development, 'a chance to make these new towns Christian communities' (Fisher, 1945a). His Final Report, recognising churches as integral to community building, proposed that around 25 sites in a town of 60,000 (four in a neighbourhood of 10,000) be reserved for places of worship on a non-profitmaking basis (MTCP, 1946). Higher than Abercrombie's estimate in the Greater London Plan of around 10–15 in a community of 60,000 (Abercrombie, 1945), both saw such provision as integral to planning new communities, although Reith was privately disappointed not to successfully promote a new form of 'union' church bringing denominations together (Fisher, 1945b).

New Towns and expanded towns

In 1943 the main Christian denominations jointly made representations for land for worship space in New Towns with Church and government records both underlining high-level commitment to planning for places of worship. Ministers agreed such site disposals at a quarter of the land's housing value (MHLG, 1951), giving assurances that churches would be consulted on their needs before development corporations submitted their overall proposals (MHLG, 1948). At a 1946 civil and ecclesiastical planning conference, a senior civil servant stated the minister would not acquiesce in any plan for New Towns not making proper provision for churches (Fisher, 1946b). The example of Stevenage was cited, its draft Masterplan incorporating provision for two churches in each of six proposed self-contained neighbourhood units. Following lobbying, Lord Silkin agreed that land for places of worship constituted 'exceptional circumstances' justifying freehold disposal (Fisher, 1946a).

Notwithstanding the lessons of the inter-war years, early New Towns often suffered a similar time lag before community facilities were provided (Ward, 1993; Church Commissioners, 1952), with significant time lapses before much other than temporary building took place, reflecting shortages in money, materials and building licences (Yelton and Salmon, 2013). A typical New Town pattern saw places of worship situated centrally within neighbourhoods but often without great architectural impact. Assumptions they would bring a sense of height and iconic distinction were largely disappointed by a prevalence of undistinguished dual-purpose buildings lacking strong visual presence (Baker, 2002), partly reflecting the challenges of funding construction.

Notwithstanding success in securing suitable, conveniently located sites on advantageous terms, concerns arose that a lack of strategic thinking led to over-provision of buildings and liturgical-only space in close proximity, including the failure to grasp ecumenical opportunities, highlighting the dilemma for the church planner: 'how to plan for mission for people who have not already arrived in the situation, without the risk of supplying something which is not suited to their needs' (Jones, 1971: 63, cited in Baker, 2002).

Denominations failing to share buildings frustrated politicians and planners. Legal obstacles were eventually overcome by the Sharing of Church Buildings Act 1970, enabling an ecumenical shared-centre model to develop. Planning consultants also advocated a more strategic approach and hierarchical system of pastoral centre at neighbourhood level (3,000–4,000), district church (12,000–15,000), and town-centre church (Nankivell, 1979, cited in Baker, 2002). Gradually, any imperative to establish presence immediately by building a place of worship receded; shared, smaller-scale, multi-use buildings were encouraged.

Although the New Town planning regime accommodated the Church of England's identified building needs, Baker (2002) identifies struggles matching the New Jerusalem rhetoric of Masterplans with realities of life on the ground, notably top-down attempts to create socially mixed neighbourhoods by compressing the intricacies of community development. Nevertheless, a subsequent government-commissioned review commended the role of faith groups in facilitating integration in New Town communities, promoting Community Development Trusts, funded by developer contributions, working with faith groups as a modern equivalent (DCLG, 2006a).

While separately, the Town Development Act 1952 promoted town expansion to relieve congestion and provide housing 'with all appropriate public services, facilities for public worship, recreation and amenity, and other requirements' (TDA, 1952: Section 1), a later study was concerned at the inadequacy of social provision and facilities in some expanded towns. Noting that Masterplans were subject to the vicissitudes of economic circumstances, insufficient early provision for community facilities suggested their less likely introduction at detailed planning stage, underlining the importance of securing sites at the outset. Places of worship did not feature prominently in the hierarchy of proposed community facilities, other than suggestions they need not be grouped with other facilities (MHLG, 1967).

New housing areas

Press reports of the Archbishop of Canterbury's comments, when visiting New Addington housing estate in 1954, on a lack of facilities for many new communities alarmed key political and planning figures concerned this would be interpreted as criticism of New Towns. Lord Silkin wrote to the Archbishop:

> I am sure you appreciate the distinction between a housing estate and a New Town: in the former case I am afraid that the speedy erection of houses has been the dominant consideration and the question of the life of

> the people occupying houses has been secondary: in the case of New Towns, however, great care has been taken to make provision for work and recreational and spiritual activities at the same time as the living accommodation has been provided.
>
> (Fisher, 1954a)

The Archbishop responded that it was not possible simply to dichotomise settlements into New Towns and housing areas:

> I was speaking in New Addington with a population of 15,000 people, a 2-shilling (10-pence) bus return fare from the middle of Croydon and back: a place of 15,000 if words mean anything is a new town. The ghastly trouble was that good intentions went astray, and it was made a pure dormitory area: it was just that which I was attacking and I am sure you will agree with me.
>
> You would like to give the title New Towns only to those where everything is done as it should be. I do not think one can avoid giving the name 'Town' to large centres of population which are clearly meant to have a life of their own: to call them merely Housing Areas is to mislead people into thinking that all they need is houses.
>
> (Fisher, 1954b)

This was reinforced in replying to Harold Macmillan:

> [T]o call it an Housing Estate glosses over the fact that it ought to be a New Town with everything a town requires, and that it has not.
>
> (Fisher, 1954c)

Between 1945 and 1972, around 3,750,000 people were rehoused in England, a further 750,000 in the pipeline. Unlike New Towns aiming to create self-sufficient communities, New Housing Areas (NHAs) did not necessarily offer a comparable level of amenities, tending towards more rapid population increase (Church Commissioners, 1987). A 1952 Church and Community Conference did not identify acquisition of places of worship sites a major problem within London but acknowledged potential difficulties for minority groups unable to apply automatically for a site in every new area but who, responding to future needs, would eventually wish to build in some (NCSS, 1952). As space becomes increasingly scarce and contested, particularly in urban areas, this becomes a major concern for newer congregations.

Statistics on early post-war church building indicate that between 1945 and 1962, 1,000 new buildings were constructed by the Church of England across the country: 408 churches, 285 dual-purpose buildings and 307 parish halls (Church Information Office, 1962). Nevertheless, having failed to keep pace with inter-war developments, the Church remained exercised by NHA challenges, facing real difficulties maintaining even the simplest point of contact with some of those rehoused. Many local authorities, with government approval, made sites available on the same advantageous terms as New Towns (MHLG, 1951) but there were concerns that the spectacle of 'Site reserved for a church' without other evidence of church life be kept to an absolute minimum with, if necessary, temporary accommodation in existing buildings (Church Assembly, 1952). Financial rather than planning issues predominated. The New Housing Areas (Church Buildings) Measure 1954 enabled the Church Commissioners to grant-aid site acquisitions and building costs in areas of major development, supporting projects in New Towns, expanded towns and NHAs, with more than 1,750 buildings grant-aided by 2000 (Church Commissioners, 2000).

1980s–1997: shifting government and church priorities

The 1980s' switch from centralist planning towards private development delivering new settlements also saw a shift towards inner-city regeneration. The *Faith in the City* report commended wider adoption of inter-denominational sharing-agreements for churches and informal sharing of other buildings with other faith groups struggling to locate urban space (Wilding, 1985). Notwithstanding inter-faith networks in areas with a strong multi-faith presence, multi-faith worship space remains highly controversial across faith groups. Planners may push for multi-faith sharing of facilities, but this remains largely unrealistic in terms of worship space itself.

The traditional Church of England ministry model of a building and priest in each parish came under increasing strain; few dioceses now start mission planning automatically intending to build a dedicated church. Many initially focus on building a sustainable church community, meeting in shared spaces such as schools or community halls (Church Commissioners, 2000). Canon law requires provision of a licensed place of worship in every parish; while an incumbent has to reside in his or her benefice for the care of all parishioners. Following concerns as the Established Church about meeting these obligations under more restrictive Green Belt development control policies, the government accepted that they could be a material planning consideration, notifying the Planning Inspectorate accordingly (DoE, 1997). Subsequent case law has primarily concerned parsonages, including the successful application of canon law as a material consideration (Appeal reference APP/G2245/A/01/1073644).

Planning policy: 1997 onwards

Despite growing public policy discourse on faith groups' contribution to civil society, national planning policy since 1997 makes little direct reference to their needs for worship space. This is usually subsumed within generic community facilities, diluting any requirement in local plans to provide specifically for worship space (ODPM, 1999) by focusing on planning for social infrastructure (DCLG, 2008). The emerging sustainability agenda brought a commitment to delivering sustainable communities meeting a diversity of needs (ODPM, 2005). Faith groups were identified as having an important part to play in neighbourhood life, in similar overtones to community building in New Towns literature (DCLG, 2006b). Alongside various political initiatives promoting faith involvement in regeneration and public policy delivery, this at least provides useful context for planning applications.

A planning policy supplement on eco-towns proposed modern versions of self-contained New Towns, requiring 'leisure, health and social care, education, retail, arts and culture, library services, sport and play facilities and community and voluntary sector facilities' without specific reference to worship facilities (DCLG, 2009: 10), but a best practice worksheet recommended mapping existing community and faith organisations and supporting early community infrastructure (including space used jointly for worship and community centres):

> On the understanding that all facilities are open to all members of the community – with or without the particular faith – developers may want to consider granting land or leases to a group to provide a community facility.
> (TCPA, 2008: 17)

A core principle of the National Planning Policy Framework (DCLG, 2012: 6) is delivering 'sufficient community and cultural facilities and services to meet local needs'; policies and decisions should 'plan positively for the provision and use of shared space, community facilities (such as local shops, meeting places, sports venues, cultural buildings, public houses, and places of worship) ... to enhance the sustainability of communities and residential environments' (2012: 17). Need for worship space is largely presented elsewhere in the context of health and well-being. Up-to-date evidence bases and community engagement are required in determining needs but accompanying guidance does not expand on places of worship, other than illustrating retention of community facilities in rural areas.

Potentially, the Localism Act 2011, including Neighbourhood Planning and Community Right to Bid, might facilitate such provision for faith groups represented on neighbourhood forums where a need is identified but the practicalities of implementation remain. The communities' secretary urged faith groups to utilise Community Right to Challenge and Community Right to Bid to strengthen their arm by transforming local buildings into, for example, centres for social enterprise (DCLG, 2011). However, this is more a call to deliver welfare services than addressing needs for worship space. More pertinent is the statutorily required pre-application community consultation for developments of a certain size, enabling faith groups to argue for space for worship.

Planning obligations

Planners may sometimes obtain more from developers, such as money towards community buildings, through what is known as 'planning gain', Section 106 agreements and planning gain mechanisms (Greed and Johnson, 2014: Chapter 3). Planning obligations or on-site provision in kind may ensure developments provide necessary community infrastructure in major housing developments. However, there is a risk, notably in an economic downturn, that such elements are sacrificed in any renegotiation of planning conditions. It is unclear whether the Community Infrastructure Levy might assist: the list of 'infrastructure' items in the Planning Act 2008 is not definitive but does not include places of worship. Even if identified in an infrastructure delivery plan, competing claims for limited funds makes assistance from this source unlikely. Many local authorities consider places of worship as community facilities without recognising worship space itself as a separate need.

Use classes

Increasing pressures on land supply and competition for space underline the significance of land-use regulation. From 1948, use for public worship was a separate use class, until a major simplification in 1987 saw this grouped with non-residential institutions under D1 use class. Disappearance as a separate use class was described as 'a further stage in the secularisation of Britain' (Home, 1992: 198); places of worship now compete with health and education centres, museums and libraries for D1 land-use allocations in local plans and Masterplans. While conversion from another D1 use to worship normally constitutes permitted development, sometimes attached conditions preclude this (Smith and Parham, 2010), exacerbating problems for expanding congregations, particularly in urban areas, facing high land values and site scarcity. Extensions to permitted development, including from office use to residential (TCPO, 2013) inhibit use of planning obligations to address community needs, placing existing facilities under greater pressure while simultaneously increasing land values and competition for space. This causes particular problems where seeking town-centre space, often the preferred place of worship location in local plans.

Equality

The Equality Act 2010 introduced a new Public Sector Equality Duty, with religion or belief one of nine protected characteristics. Its potential impact, for example in challenging provisions in local plans or planning decisions, is still unclear although in London the importance of addressing increasing demand for worship space for a growing and diverse population has been underlined (GLA, 2013); while the London plan itself includes places of worship within the definition of social infrastructure (GLA, 2011). Concerns that the understanding of need were based on an old-fashioned Church of England model not adequately reflecting those with different patterns of worship (GLA, 2007) were understood as meaning that a geographical parish model does not encompass 'gathered' churches with wide catchments requiring large spaces (Rogers, 2013).

Local plans

Some local plans include specific policies on places of worship but for many these fall within a broader definition of community/social infrastructure.

There are examples, notably in urban areas with significant multi-faith presence, of special planning guidance, sometimes produced following high incidences of planning refusals or illegal use of unsuitable premises (BCC, 2011). Such guidance may rehearse potential problems as much as address them, focusing on compliance and external impacts (Gale and Naylor, 2002) including those arising from customary religious practices such as the Muslim call to prayer (Peach and Gale, 2003). Policies include identifying residential areas as unsuitable for activities attracting large numbers, demonstrating that predominantly local needs are being met, without adverse effect on neighbours (LBH, 2003); concerns that designs sit harmoniously with surroundings (BCC, 2011); and use of planning conditions to limit operational times, restricting activities and use of public address systems (London Borough of Havering, 2014; LBH, 2003).

Specially commissioned need studies

Various specially commissioned needs surveys in recent years included one scrutinising all London planning applications and appeals since 2000, concluding that while finding worship space in London was an historic problem, many faith groups now experienced significant difficulties, partly due to the operation of the planning system and inadequate provision of D1 land, with poor communication and a lack of 'faith literacy' among many planning authorities. It recommended specific planning policies and guidance on assessing and meeting needs, including use of planning conditions and providing land or space for faith groups in Design Briefs and Masterplans, suggesting a 'standards'-based approach in new development areas with, as a possible benchmark, setting aside 0.5 hectares of land per 3,000 new population (CAG, 2008a; 2008b).

Such studies occur mainly in growth areas or those with significant multi-faith presence. Cambridgeshire's local authorities were recommended to adopt a similar standard policy of 0.5 hectares of free land (or equivalent in buildings and/or financial contribution) for each 3,000 dwellings in new communities, on the basis that faith groups would provide a wide range of community facilities in perpetuity, while enjoying a 'private' area for worship. The rationale was based on experience at Cambourne and Milton Keynes, suggesting smaller sites would provide inadequate space to serve growing congregations and support wider facilities. Recognising that some groups needed larger sites (particularly minority faiths drawing members from a wider area), adequate D1 land should be zoned to enable faith and other community groups to access land at an affordable price; planning policy should also allow use of employment land providing other relevant planning considerations were met (Three Dragons, 2008). A Thurrock study mapped all faith premises by factors such as attendance, building condition and catchment area, calling for clear policy criteria for determining new, or converting existing, premises and flexibility on releasing employment land for larger spaces. Other proposals included maintaining databases of faith groups and their premises and faith forums for ongoing engagement (Buchanan, 2009).

Emphasised elsewhere were the importance of up-to-date evidence bases to demonstrate need, specific places of worship policies in development plans, linked to Core Strategy policy on delivering community regeneration. Consideration should be given to locational criteria, applying general design principles flexibly with regard to certain types of worship and use of mitigating planning conditions (such as sound-proofing, restricted hours of operation, green travel plans and parking provision) (CAG, 2013).

Other reviews highlight the need for adequate, reliable funding, linking lack of community facilities with unfulfilled Section 106 agreements, advocating enforcement of such agreements and setting land aside for future community buildings to ensure adequate provision (Stott et al., 2009). The Joseph Rowntree Foundation commended English Partnerships' approach of carrying out demographic projections, including for faith facilities, to underpin service-provider planning (Taylor, 2011).

Delivering social infrastructure in modern major housing developments is challenging but with worship space increasingly defined as social or community infrastructure, grouped alongside health, education and leisure needs, it struggles to be recognised as a specific, standalone need. Draft planning guidance for London (GLA, 2014: 84) identified faith buildings as 'fundamental to the creation of Lifetime Neighbourhoods', calling on boroughs to engage with principal faith groups to identify their requirements in local and strategic assessments, planning positively for provision of shared space to enhance community sustainability. Noting evidence that the number of D1 use buildings was diminishing, it proposed identifying suitable locations in Opportunity and Intensification Areas and large residential developments, and not adopting planning conditions preventing D1 use buildings being used for worship.

Diocesan practices and networks

The Church of England has no set criteria defining the level of development necessitating a place of worship, or at what stage. The ecumenical Churches Together in New Housing Areas adopts '1,000 new homes' as the unit large enough to raise questions about providing facilities, factors informing such decisions including local plan provision, development drivers, population profile, housing approvals, economic circumstances and proximity to, and capacity of, other church buildings. London Diocese has mapped all new housing development plans and identified several strategic priorities, based largely on the London Opportunity Areas. It works with local parishes to identify and prioritise needs, so that once potential D1 use sites are identified, it can negotiate with developers and planners.

Case examples

Taking examples of recent (at the time) high-profile housing developments on rural greenfield and urban brownfield sites highlights the interaction between various strands of planning policy and practice.

Cambourne Church

In 1994, outline consent was granted for the new settlement of Cambourne, near Cambridge, comprising 3,300 dwellings and associated facilities. This included provision that by completion of the 500th dwelling, the owners should offer for nil consideration, 'to a relevant organisation to provide on the Site one acre or thereabouts of serviced land for the construction of an ecumenical centre' (SCDC, 1994: 45). Separately a community centre would be provided by completion of the 1,000th dwelling. The 1995 Cambourne Masterplan included prime space for a church on High Street.

In 1999 the first residents moved in and the Churches' Ecumenical Council appointed a minister to establish a community church. Worshippers initially met in the doctor's waiting room but in 2001 gained planning consent for temporary Portakabin accommodation, The Ark (SCDC, 2001), on land let by the developers on a year-on-year rolling licence (SCDC, 2013). This was the first, belated, community space in Cambourne, underlining difficulties community groups faced sharing the same space following failure to meet milestones in the planning agreement and late arrival of facilities (Stott et al., 2009; Platt, 2007). In 2002 the developers transferred the church site to Shared Churches (Ely) Ltd in accordance with the planning agreement. By 2005 the congregation was worshipping in the newly provided community hall while fundraising to construct the permanent ecumenical church, opened in 2009 (with café and letting space).

Fundraising continued for a church extension. When housing expansion was planned in 2010, Cambourne Church sought a contribution to meet additional need for religious and community facilities, resulting in a significant financial contribution (SCDC, 2010) but community facilities remained under pressure, with over 300 groups competing for only six meeting spaces by 2012. The demography also proved more complex and diverse than anticipated with a significant Muslim community at the point of needing its own facility (Cearns, 2012).

Barking Rivergate Centre

The Barking Local Plan included provision of schools, religious buildings, welfare and associated public service facilities among requirements for comprehensive development of Barking Reach (LBBD, 1995). Bellway Homes acquired Barking Riverside, a 179.34-acre brownfield site identified for residential-led development, building 900 new homes by 2000 but providing little social or transport infrastructure.

Rather than piecemeal progress, the local authority and English Partnerships saw potential to deliver higher quality development through masterplanning the whole site and closer partnership with the developer (NAO, 2007). In 2004, Barking Riverside Limited was established as a joint public/private venture between English Partnerships and Bellway Homes. An overall Masterplan was produced and outline permission granted in 2007 for a 20-year programme (including 10,800 residential dwellings, new schools, commercial, leisure and community uses). Based on four development stages, this provided for three new neighbourhood centres and a district centre.

A planning agreement required not more than 1,000 occupied units in Stage 1 without providing a place of worship, children's centre and community facility within a neighbourhood centre; similar facilities linked to occupancy levels were earmarked for the remaining phases (LBBD, 2007). Site-specific allocations outlined careful phasing to community infrastructure matched to increased housing, with each new centre to have community hubs (schools, religious meeting places and other social infrastructure) (LBBD, 2010b). This aligned with Core Strategy policy whereby major new developments should provide or contribute towards any additional need arising for community facilities, phased to ensure their availability as new developments became occupied, encouraging dual use of premises for a range of community uses (LBBD, 2010a).

Detailed planning permission for the first stage was granted in 2010 (LTGDC, 2010). Construction of the community hub commenced shortly before work began on the initial 357 homes, ensuring facilities were in place as residents arrived. The local authority worked through the local faith forum to identify which worshipping community should be offered this initial space. A joint bid from the Church of England, Methodists and Salvation Army was successful and the resulting Rivergate Centre (opened 2011) comprised a primary school, nursery, Rivergate Church, café and community facilities, co-locating a range of complementary uses within a community hub.

Tottenham Hale

Haringey's Local Plan (LBH, 2006b) identified good community facilities, including places of worship, as essential to a successful and inclusive society, recognising that supporting religious groups finding suitable premises meant residents benefiting from services provided alongside worship (LBH, 2003).

A residential-led Masterplan for a 4.8-hectare brownfield site, part of the wider Tottenham Hale Opportunity Area, focused primarily, in terms of community facilities, on providing educational and public space (LBH, 2006a). In 2007 the Hale Village Masterplan obtained outline consent for major, mixed-use regeneration, housing 16,000 new residents by 2025, with offices, retail, health centre and primary school (the latter provided through a planning agreement).

In the economic downturn, development did not immediately proceed. Meanwhile, concerned at lack of provision for any church or community facility,

London Diocese began discussions with Haringey and the developers. Since the council no longer needed a school on site, a revised planning agreement provided a significant contribution towards constructing a community centre shell (LBH, 2012a). Revised plans for the school site provided for a community centre, including nursery and café use and multi-purpose hall, within a seven-storey residential block. The developer constructed the centre to a 'shell and core standard', leased to the diocese to fit out, staff, manage and operate the whole community centre space.

Initially the planning committee deferred its consideration to clarify the proposed use of the community centre, particularly the extent of use for religious purposes (LBH, 2012a; Lee Valley Estates, 2012). A joint developer/diocesan statement outlined the latter's commitment of substantial resources to provide greater social and community infrastructure within Tottenham Hale. The building's primary use would be as a community facility operating on an open and inclusive basis, with religious use only a small part of the overall activities (the hall being used for worship on appropriate days and for special services but available at all other times to support a wide range of secular community activities). Satisfied this would be a flexible community facility supporting Hale Village, the committee granted planning permission (LBH, 2012b).

Cranbrook

Proposals emerged in the early 1990s in the Devon structure plan for a greenfield mixed-use new community. The local plan subsequently provided, initially, for 2,900 housing units on the preferred site by 2016 (EDDC, 2006a). In 2002, ten years before the first houses were built, Churches Together in Devon (CTID) formed an ecumenical planning group, participating in a council-led community working group on planning guidance to include a place of worship among the requisite community facilities (EDDC, 2005). CTID regarded a resident minister as the most effective means of delivering early community ties within the settlement; the physical church structure could come later. Following its input, the local plan inquiry inspector recommended a modification, including reference to providing dwellings for resident workers alongside places of worship, subsequently included in the plan. The economic downturn delayed progress until 2010 when outline permission was granted for Cranbrook New Community (EDDC, 2006b).

Working in partnership with the council, CTID secured, through a planning agreement, a dwelling house for a minister, use of a community centre and a serviced site for a place of worship in a strategic location (to be funded and built by CTID at some point in the medium to long term) (EDDC, 2011). Although CTID did not seek a building at this stage, the developer agreed to provide 0.5 hectares of land to 'draw down' for £1 for a place of worship, made available and transferred to the relevant authority prior to occupation of 500 dwellings, while the house for the resident church worker would be transferred prior to occupation of 150 dwellings. CTID also secured office space within the community hall built in the first phase. The church minister was in place by 2012, when the first residents arrived (Church Commissioners, 2013) using the new primary school as an interim place of worship.

These examples illustrate various routes to providing worship space in new developments, including use of Masterplans and planning agreements linked to phases of housing completion or occupation. However, such arrangements are vulnerable to economic downturns and potential renegotiation of planning agreements and may necessitate major financial input by the church body, difficult to replicate on a large scale. The two greenfield examples, Camborne and Cranbrook, most approximate the New Town experience of securing sites but postponing building churches until funds were raised or it was deemed timely.

Conclusions

The professionally led idealism and social optimism dominating post-war planning lay considerable emphasis on developing new, self-sufficient communities. Need for places of worship was recognised in public policy, underlining commonality in the goals and ideals of New Town planners and the Church of England's understanding of its role helping develop new communities (Baker, 2002). The scale of the centralist New Towns programme and availability of greenfield land meant such needs could be readily accommodated in planning terms. Government agreed site disposals on advantageous terms while Masterplans made specific provision for places of worship. More logistically challenging was the rapid population growth elsewhere, in part reflecting mass migration from inner cities. While evidence suggests that often sites for churches were also made available on advantageous terms, logistical problems and fundraising could mean lengthy delays before construction.

Fast forwarding to the present, planning for large-scale residential development now operates within a very different planning, funding and governance regime. Primarily market-led development is often undertaken piecemeal, increasingly without benefit of public investment. Limited access to major greenfield development results in severe constraints on land availability. The modern case examples illustrate the complexities of ensuring adequate community infrastructure, including places of worship. For the Church of England, with much time and effort there has been some success in securing suitable sites, including through planning obligations; cumulative population growth through smaller housing developments makes such provision more problematic. Achieving satisfactory outcomes relies heavily on relationship-building with planners and developers and prioritising needs and these examples highlight strategic commitment of significant but limited Church resources in high-profile developments, not necessarily replicable elsewhere. Where feasible, ecumenical partnerships and greater sharing of buildings helps optimise resources.

The Church of England's approach to providing worship space reflects evolving internal debate on building and supporting new communities. Sharing buildings has become more prevalent, with greater discernment on whether and when to build a dedicated place of worship rather than utilise other available venues. Even if building may not be an immediate priority, pre-emptive site allocations, as at Cambourne and Cranbrook, reserve longer-term options for a growing congregation needing its own building. Initial efforts may concentrate on community building by deploying a minister or lay worker (CCU, 2011), or providing a church school, which can both be attractive to developers and enable wider use for worship space (Church Commissioners, 2011). The Cranbrook initiative in securing a minister's residence was mutually beneficial, supporting an early Church presence and encouraging community building. Similarly, multi-use places of worship now focus on flexible space, possibly under leasehold, further breaking down barriers between sacred and secular. Indeed, multi-use facilities are often presumed as the norm in local plans. Identifying gaps in community development, such as at Haringey, is part of the Church's response both to serving the needs of people with no prior church connection and as the Established Church within a religiously plural society but requires active engagement with local plans and development proposals and deployment of specialist planning resources.

While the Church of England, with around 16,000 church buildings in established localities, may focus on providing facilities within newly created communities, other denominations and faith groups often face very different challenges, particularly large diaspora communities seeking suitable premises for growing congregations. Increasing pressure on contested urban space and high land values make it difficult for many to meet their needs and local plans can seem more constraining than facilitating in such circumstances.

Various recent needs studies propose adoption of space standards, based largely on population size (echoing Reith and Abercrombie), but compared to planning for education and health, predicting worship space needs is problematic. Faith groups differ on the size of community for which dedicated worship space is required and catchment areas, particularly for gathered congregations, can extend well beyond immediate neighbourhoods. Needs studies relate different levels of participation and visibility, especially for congregations less familiar with planning structures, while predicting future community profiles for New Housing Areas is not straightforward. Even so, worship space needs are often not directly addressed in the planning system, notwithstanding support and encouragement in public policy discourse for faith-based organisations as providers of services with wide community benefit. Unless local plans specifically recognise worship space need, any prospect of this being met through means such as planning obligations diminishes. The conflation of places of worship within the same D1 use class as a broad range of community and education uses requires faith groups to compete with many other interests for a limited and often inadequate supply of land, resulting in use of unsuitable premises and recourse to enforcement. Focusing on regulatory aspects (such as traffic, parking and noise), including restrictions on using existing D1 land for worship use, may outweigh attempts to resolve unmet need; planning practice can be problematic, even where policy is more positive.

Looking ahead, more needs-related evidence studies or faith group audits would help identify and quantify the scale of need for worship space, with space quotas or standards linked to population size offering one way forward. Leicester City Council's monitoring arrangements for planning applications include a pro-active registration system, helping with site searches and promoting a better understanding of needs; dedicated faith group planning officers can similarly assist, while strategic planning policy should address under-provision and have regard to the level and consistency of demand (ODPM, 2006). Given the tendency to subsume places of worship within 'community' infrastructure, planning obligations supporting multi-purpose facilities should be encouraged. However, in promoting co-location and shared worship facilities, the practicalities and sensitivities around possible multi-faith worship space need to be recognised. Possible locations could be identified when preparing site allocations and Masterplans; similarly, employment land reviews might look to identify suitable sites by promoting mixed uses in industrial areas and business parks.

Faith groups, ranging from large, nationally organised bodies with very significant membership to small congregations with no formal associations, cover a broad spectrum of need, from large purpose-built structures to hiring small meeting rooms. They host a growing variety of community and welfare functions, with food banks and credit unions among more recent additions, often relieving statutory service provision. This role, espoused in public policy discourse, is endangered where the planning system fails to meet their accommodation needs which include worship space.

References

Abercrombie, P. (1945) *Greater London Plan*, London: HMSO.

Baker, C. (2002) *Towards a Theology of New Towns: The Implications of the New Towns Experience for Urban Theology*. Unpublished PhD thesis, Manchester: University of Manchester.

BCC – Birmingham City Council (2011) *Places of Worship and Faith-related Community and Educational Uses*, The Birmingham Plan, Supplementary Planning Document, Birmingham City Council.

Brierley, P. (2016) *UK Church Statistics 2005 to 2015*, Tonbridge: ADBC Publishers.

Buchanan, C. (2009) *Provision of Facilities for Faith Groups in Thurrock: Final Report*, Grays: Thurrock Council.

Burgess, R. (2009) 'African Pentecostal spirituality and civic engagement: the case of the Redeemed Christian Church of God in Britain', *Journal of Beliefs and Values: Studies in Religion and Education* 30(3): 255–273.

CAG Consultants (2008a) *Responding to the Needs of Faith Communities: Places of Worship: Evidence Report Accompanying the Final Report to Greater London Authority*, London: CAG.

CAG Consultants (2008b) *Responding to the Needs of Faith Communities: Places of Worship: Final Report to Greater London Authority*, London: CAG.

CAG Consultants (2013) *Northampton Faith Communities and Places of Worship Audit and Needs Assessment: Final Report to Northampton Borough Council*, London: CAG.

CAG Consultants (2017) *Faith Groups and Meeting Places: Evidence Base Study*, London: London Borough of Barking and Dagenham.

Cearns, E. (2012) *Faith in New Developments: Partnerships, People and Places*, Cambridge: East of England Faiths Council.

Chandler, A. (2006) *The Church of England in the Twentieth Century: The Church Commissioners and the Politics of Reform, 1948–1998*, Woodbridge: Boydell Press.

Chapman, R. and Lowndes, V. (2008) 'Faith in governance? The potential and pitfalls of involving faith groups', *Urban Governance, Planning Practice and Research*, 23(1): 57–75.

Church Assembly (1944) *The Church and the Planning of Britain: Report of the Social and Industrial Commission*, CA 753, London: Church Assembly.

Church Assembly (1952) *The Church in New Housing Areas: Report of the Committee on The Church in New Housing Areas*, CA 1024, London: Church Assembly.

Church Commissioners (1952) Estates and Finance Committee Documents and Papers for Diocesan Conference on New Housing Areas in CC/SEC/PLAN/5. London: Church of England Records Centre.

Church Commissioners (1987) *Report to the Pastoral Committee on the New Housing Areas Grant Scheme*, London: Church Commissioners.

Church Commissioners (2000) *Report to the Pastoral Committee on Grants for Church Buildings*, London: Church Commissioners.

Church Commissioners (2011) *Resourcing Mission Bulletin: The Church's Mission in New Housing and Other Development Areas*. www.churchofengland.org/media/1268347/resourcing%20mission%20bulletin%20-%20may%202011.pdf [accessed 31 May 2019].

Church Commissioners (2013) *Church Growth Resourcing Mission Bulletin Pioneer Ministry in a New Community: A Case Study on the First Year of Ministry in Cranbrook, Exeter*. www.churchgrowthrd.org.uk/UserFiles/File/Resourcing_Mission_Bulletin/Oct_2013/5._Pioneer_ministry_in_a_new_community.pdf [accessed 2 July 2014].

Church Information Office (1962) 'Travelling exhibition on post-war church building', Press notice, 12 December, London: Church Information Office.

Comedia (1995) *Spirit of Place: Redundant Churches as Urban Resources*, Stroud: Comedia.

CCU – Council for Christian Unity (2011) *Fresh Directions in Unity in Mission. Part 2:D. Unity in Mission in New Housing Areas*, CCU Paper (2011)10. www.cte.org.uk/Articles/317764/Home/Resources/New_Housing_Areas/Church_and_New.aspx [accessed 26 July 2014].

Cullingworth, B. and Nadin, V. (2006) *Town and Country Planning in the UK*, London: Routledge (14th edition).

Davie, Grace (2015) *Religion in Britain: A Persistent Paradox*, London: Wiley.

Deakin, N. (2010) 'Religion, state and third sector in England', *Journal of Political Ideologies* 15(3): 303–315.

DCLG – Department for Communities and Local Government (2006a) *Transferable Lessons from the New Towns, Housing*, Oxford: Department of Planning Oxford Brookes University.

DCLG – Department for Communities and Local Government (2006b) *Strong and Prosperous Communities*, The Local Government White Paper, London: The Stationery Office.

DCLG – Department for Communities and Local Government (2008) *Planning Policy Statement 12: Creating Strong, Safe and Prosperous Communities through Local Spatial Planning*, London: The Stationery Office.

DCLG – Department for Communities and Local Government (2009) *Planning Policy Statement: Eco-towns*, A supplement to Planning Policy Statement 1, London: The Stationery Office.

DCLG – Department for Communities and Local Government (2011) 'New powers will put faith groups at the centre of their communities', Press notice, 16 June 2011.

DCLG – Department for Communities and Local Government (2012) *National Planning Policy Framework*, London: The Stationery Office.

DoE – Department of the Environment (1997) 'Material planning consideration: Church of England canon law', Letter from the Planning Directorate to all Planning Officers, 4 June.

Eade, J. (2011) 'From race to religion: multiculturalism and contested urban space' in J. Beaumont and C. Baker (eds) *Postsecular Cities: Space, Theory and Practice*, London: Continuum, 154–167.

EDDC – East Devon District Council (2005) *Development Control Committee Main Report on Proposal for Dev7/22/03/P1900/00791* (OUTLINE). www.eastdevon.gov.uk/plg_new-comm-report-maintext.pdf [accessed 14 June].

EDDC – East Devon District Council (2006a) *Adopted East Devon Local Plan 1995 to 2011*, Honiton, Devon: East Devon District Council.

EDDC – East Devon District Council (2006b) *Supplementary Development Control Committee Report on 7/22/03/P1900/00791 for 19th December 2006*. www.eastdevon.gov.uk/plg_cranbrook-latestreport.pdf.

EDDC – East Devon District Council (2011) s106 agreement Reference 03/P1900, Schedule 22. http://planningapps.eastdevon.gov.uk/Planning/StreamDocPage/obj.pdf?DocNo=809927&PageNo=1&content=obj.pdf.

Farnell, R., Furbey, R., Shams, A.-H., Hills, S., et al. (2003) *'Faith' in Urban Regeneration? Engaging Faith Communities in Urban Regeneration*, Joseph Rowntree Foundation, Bristol: Policy Press.

Fisher, G. (1945a) Letter to the Archbishop of Canterbury from Lord Reith. Box 49, Folder 290. Archbishop Geoffrey Fisher Papers, Lambeth Palace Library.

Fisher, G. (1945b) Letter from the Archbishop of Canterbury to Sir James Brown, Third Church Estates Commissioner. Box 49, Folder 290E. Archbishop Geoffrey Fisher Papers, Lambeth Policy Library.

Fisher, G. (1946a) Copy of letter from Lord Silkin to the Bishop of London dated 10th October 1946. Box 49, Folder 332. Archbishop Geoffrey Fisher Papers, Lambeth Policy Library.

Fisher, G. (1946b) Note of a Conference on Ecclesiastical and Civil Replannng dated 8th October 1946. Box 49, Folder 337. Archbishop Geoffrey Fisher Papers, Lambeth Policy Library.

Fisher, G. (1954a) Letter from Lord Silkin, dated 26th March 1954. Box 145, Folder 210. Archbishop Geoffrey Fisher Papers, Lambeth Palace Library.

Fisher, G. (1954b) Letter from Archbishop of Canterbury to Lord Silkin, dated 26th March 1954. Box 145, Folder 211. Archbishop Geoffrey Fisher Papers, Lambeth Palace Library.

Fisher, G. (1954c) Letter from Archbishop of Canterbury to Rt Honourable Harold Macmillan, dated 27th March 1954. Box 145, Folder 213. Archbishop Geoffrey Fisher Papers, Lambeth Palace Library.

Gale, R. (2008) 'Locating religion in urban planning: beyond "race" and ethnicity?', *Planning Practice and Research* 23(1): 19–39.

Gale., R. and Naylor, S. (2002) 'Religion, planning and the city: The spatial politics of ethnic minority expression in British cities and towns', *Ethnicities* 2: 387.

Garbin, D. (2013) 'The visibility and invisibility of migrant faith in the city: diaspora religion and the politics of emplacement of Afro-Christian churches', *Journal of Ethnic and Migration Studies* 39(5): 677–696.

Ghatak, S. and Abel, A. (2013) 'Power/faith: governmentality, religion, and post-secular societies', *International Journal of Politics, Culture and Society* 26: 217–235.

GLA – Greater London Authority (2007) *Planning for Equality and Diversity in London: Supplementary Planning Guidance to the London Plan*, Mayor of London, London: GLA.

GLA – Greater London Authority (2011) *The London Plan: Spatial Development Strategy for Greater London*, Mayor of London, London: GLA.

GLA – Greater London Authority (2013) *Assessment of the GLA's Impact on Faith Equality*, London: GLA.

GLA – Greater London Authority (2014) *Draft Social Infrastructure Supplementary Planning Guidance: London Plan 2011 Implementation Framework*, Mayor of London, London: GLA.

Greed, C. (2011) 'A feminist critique of the postsecular city: god and gender' in J. Beaumont and C. Baker (eds) *Postsecular Cities: Space, Theory and Practice*, London: Continuum, 104–119.

Greed, C. (2016) *Christianity and Planning: Pentecostal Churches*, Occasional Paper 1, Bristol: University of the West of England.

Greed, C. with Johnson, D. (2014) *Planning in the United Kingdom: An Introduction*, London: Palgrave Macmillan.

Home, R. (1992) 'The evolution of the use classes order', *Town Planning Review* 63(2): 187–201.

Jones, D. (1971) *Planning for Change*, A New Town Ministers Association Publication.

Lee Valley Estates (2012) 'Church-run community centre and flats approved for Tottenham's Hale Village', Press release, 8 July. www.leevalleyestates.co.uk/news_detail.html?id=129 [accessed 2 July 2014].

LBBD – London Borough of Barking and Dagenham (1995) *Unitary Development Plan*.

LBBD – London Borough of Barking and Dagenham (2007) Summary of main planning obligations in S.106 Agreement signed 7 August 2007. Barking Riverside planning application 04/01230/OUT. www.lbbd.gov.uk/Environment/PlanningPolicy/Documents/s106AgreementBarkingRiverside.pdf [accessed 26 June 2014].

LBBD – London Borough of Barking and Dagenham (2010a) *Core Strategy. Planning for the Future of Barking and Dagenham.*

LBBD – London Borough of Barking and Dagenham (2010b) *Planning for the Future of Barking and Dagenham: Site Specific Allocations Development Plan Document.*

LBH – London Borough of Haringey (2003) *Special Planning Guidance 11b – Buildings Suitable for Community Use.*

LBH – London Borough of Haringey (2006a) *Transforming Tottenham Hale. Urban Centre Masterplan: Supplementary Planning Document.* www.haringey.gov.uk/tottenham_hale_spd_low_res.pdf [accessed 2 July 2014].

LBH – London Borough of Haringey (2006b) *Haringey Unitary Development Plan.*

LBH – London Borough of Haringey (2012a) Officers' Report for consideration at Planning Sub-Committee, 28 June 2012, on former GLS Depot, Ferry Lane, Tottenham N17.

LBH – London Borough of Haringey (2012b) Minutes of a Meeting of the Planning Sub-Committee, 28 June 2012, on former GLS Depot, Ferry Lane, Tottenham N17.

London Borough of Havering (2014) *Planning Guidance for Faith Groups*. www.havering.gov.uk/Documents/Planning/planning-requirements-faith-groups.pdf [accessed 27 July 2014].

LTGDC – London Thames Gateway Development Corporation (2010) Planning Committee Meeting on 8 February 2010. Planning Application Report of the Director of Planning. Planning Application for Determination by the London Thames Gateway Development Corporation (LTGDC-09-107-FUL).

Mansfield, J. (2007) 'Evolving heritage control and practice: the case of Anglican churches in English parishes', *Structural Survey* 25(3/4): 265–278.

MHLG – Ministry of Housing and Local Government (1948) Briefing Note to Parliamentary Secretary dated 4th February 1948 in iHL/90/343 Churches and Chapels. London: The National Archives.

MHLG – Ministry of Housing and Local Government (1951) in HLG/90/343 Churches and Chapels. London: The National Archives.

MHLG – Ministry of Housing and Local Government (1967) *The Needs of New Communities: A Report on Social Provision in New and Expanding Communities*, London: HMSO.

MTCP – Ministry of Town and Country Planning (1946) *Final Report of the New Towns Committee*, Cmd.6876, London: HMSO.

Mulkeen, Andrea (2014) *The Planning System and Places of Worship: The Extent to Which the Planning System since the Second World War Has Facilitated the Needs of Faith Communities for Places of Worship*, London: University of Westminster.

Mynors, C. (1985) '"Render unto Caesar ...": the ecclesiastical exemption from listed building control', *Journal of Planning and Environment Law*, pp. 599–610.

Nankivell, C. (1979) *Religion in the New Town*, Unpublished MSc Thesis, Birmingham: University of Birmingham.

NAO – National Audit Office (2007) *The Thames Gateway: Laying the Foundations. Report by the Comptroller and Auditor General.*

NCSS – National Council of Social Service (1952) *Church and Community in the New Housing Areas. Report of Central Churches Group.*

ODPM – Office of the Deputy Prime Minister (1999) *Planning Policy Guidance 12: Development Plans*, London: HMSO.

ODPM – Office of the Deputy Prime Minister (2005) *Planning Policy Statement 1: Delivering Sustainable Development*, London: HMSO.

ODPM – Office of the Deputy Prime Minister (2006) *Review of the Evidence Base on Faith Communities*, London: HMSO.

Osborn, F. and Whittick, A. (1977) *New Towns: Their Origins, Achievements and Progress*, London: L. Hill.

Peach, C. and Gale, R. (2003) 'Muslims, Hindus, and Sikhs in the new religious landscape of England', *The Geographical Review* 93(4): 463–490.

Platt, S. (2007) *Lessons from Cambourne*, Cambridge Architectural Research Limited.

Rogers, A. (2013) *Being Built Together: A Story of New Black Majority Churches in the London Borough of Southwark*, London: University of Roehampton.

Rogers, A. and Gale, R. (2015) *Faith Groups and the Planning System (England): Policy Briefing to Parliament*, London: University of Roehampton.

Smith, P. and Parham, S. (2010) 'Making space for worship', *Planning Resource*. www.planningresource.co.uk/article/983094/making-space-worship [accessed 2 February 2014].

SCDC – South Cambridgeshire District Council (1994) Section 106 Agreement, in relation to planning application S/1371/92/0 dated 20th April 1994. http://plan.scambs.gov.uk/swiftlg/MediaTemp/1118883-165205.pdf [accessed 2 July 2014].

SCDC – South Cambridgeshire District Council (2001) Planning Application S/6128/01/F Decision Notice.

SCDC – South Cambridgeshire District Council (2010) Report to Planning Committee on 6th December 2010 on Planning Application S/6438/07/O.

SCDC – South Cambridgeshire District Council (2013) Delegation Report on Planning Application S/1051/13/VC.

Stott, M., Stott, N. and Wiles, C. (2009) *Learning from the Past? Building Community in New Towns, Growth Areas and New Communities*, Thetford: Keystone Development Trust.

Taylor, M. (2011) *Community Infrastructure in New Residential Developments*: York: Joseph Rowntree Foundation.

Three Dragons (2008) *Facilities for Faith Communities in New Developments in the Cambridge Sub-Region: Report for Cambridgeshire Horizons.*

TCPA – Town and Country Planning Association (2008) *Towards Sustainable Communities: Eco-towns Community Worksheet. Advice to Promoters and Planners.*

Town and Country Planning (General Permitted Development) (Amendment) (England) Order (TCPO) 2013.

Ward, C. (1993) *New Town, Home Town*, London: Calouste Gulbenkian Foundation.

Wilding, P. (1985) *Faith in the City: The Report of the Archbishop of Canterbury's Commission on Urban Priority Areas*, London: Church House Publishing.

Yelton, M. and Salmon, J. (2013) *Anglican Church-Building in London 1946–2012*, Reading: Spire Books.

13 A woman faith practitioner's experience in the UK night time economy

Gill Reeve

Abstract

The case study in this chapter is based upon my experience as a faith practitioner, who set up an inclusive urban sacred space, namely a night time church within the context of the night time economy of a medium-sized city in the northwest of England. The account is informed by three factors, concepts of the night time economy, networked urban governance theory and introducing faith into the urban public sphere. The study also draws on assemblage thinking and a Foucauldian understanding of power relations to cast light upon the complex relational web that was found to form around the creation of 'Night Church'. Alliance with, and the involvement of, a range of other governmental and religious organisations was found either to enable and sustain this faith practice or to create difficulties. In particular, the gendered nature of contested power relations with other, more established faith institutions and their male leadership proved problematic. As an auto-ethnographic study, the findings are person-specific and contextual. But they provide new insights into the under-researched field of the relationship between faith practice and urban governance and how faith actors are negotiating complex and multiple power relations in the public sphere. Therefore, relatively speaking the emphasis is more upon the networks that constrained the running of the night church, rather than upon the characteristics of the church and its members.

Key words: night time e conomy, faith practice, networked urban governance, night church

Introduction and context

This case study explores my experience as a woman faith practitioner, leading an inclusive, Christian sacred space, namely a 'Night Church' in the heart of a northwest UK city within the context of the night time economy (NTE). The NTE emerged in the late-1980s, as part of a wider, optimistic vision of '24 hour cities' (Roberts and Eldridge, 2009). The contemporary NTE has been described as 'the assemblage of bars, clubs, cinemas, theatres and cultural festivals and events at night time which are, in a context of urban entrepreneurialism, supposed to contribute to urban regeneration and local economic growth' (van Liempt et al., 2015: 412; Hall and Hubbard, 1998). Within the NTE, a mixture of formal and informal partnerships and networks were to be found, bringing together a variety of actors with differing narratives, motivations and perspectives that shape discussion, decision making and collaborative action (Dinham and Lowndes, 2009). Faith organisations and faith actors have become increasingly visible within these networks and partnerships, including, national faith-based organisations (FBOs) such as Street Pastors and smaller, local faith initiatives such as my 'Night Church'. The case study adopts the term 'faith' throughout because in governance working the term 'faith' is used in preference to 'religion' and has come to represent the main religious traditions recognised by government (Dinham, 2017). I started Night Church in 2010 and led the project till 2016 when it was handed over to members of a leadership team, when I began ordination training within the Anglican Church. It still flourishes with up to 200 people visiting the Night Church and a new leadership team comprising around ten people, taking charge. All locations, organisations and individuals referred to in this chapter are anonymised. It was important for me to observe how my faith practice was being drawn into the web of what may be termed 'networked urban governance' (NUG). Therefore, relatively speaking the emphasis in this chapter is more upon the networks that constrained the running of the Night Church, rather than upon the characteristics of the church and its members.

Conceptual approach: relational entanglements in networked urban governance

Networks and networking

My case study is auto-ethnographic and acknowledges that 'an individual's knowledge is partial' and that there is no such thing as 'raw experience' since all experience is mediated by interpretation (Maynard and Purvis, 1994: 6). This case study analyses three distinct but interrelated parts of my networks in running the church; NTE relationships, urban governance relationships and relationships within the local faith network. These relational webs variously enabled or constrained my 'faith practice' (that is my personal 'praxis' of putting faith into practice by setting up the Night Church). This case study shows how my faith experiences in setting up the Night Church have been shaped by changes to local government arising from the neoliberal, social policy agenda of the late twentieth century (Pierre and Peters, 2000). This change in agenda, and the related government funding cutbacks, has resulted in local government being increasingly dependent on other social actors in the policy formulation and delivery of community and welfare services, especially the church. The three key factors that were framed by my study, namely the NTE, NUG and the role of faith in the public sphere, will now be discussed in more detail.

The night time economy

Late-night cities are diverse and complex, shaped by an 'intertwined array of historical, local and global, economic and social trajectories' (Roberts and Eldridge, 2009: 10). Cities have always had a distinctive social imaginary from the daytime, often conceived as liminal places of pleasure or danger (Roberts and Eldridge, 2009). Concerns over security, morality and accessibility are constantly recurring themes, with the conflict between the desire for freedom and the desire for order in constant tension (Schlor, 1988). The city at night is widely perceived as an alternative time-space that is less hurried (Roberts and Eldridge, 2009) and with greater commonality and sociability, contrasting it from the busyness and anonymity of the daytime (Melbin, 1978). But the familiar, historic trope of the (male) *flâneur* and the (female) prostitute also points to the gendered nature of cities at night (Schlor, 1988), shaping how women experience and participate in the urban night. Research has often focused on the absence of women and the oppressive and excluding nature of the late-night city for women. But more recent research has demonstrated that women have been active within the urban night and has provided new opportunities and freedoms (Roberts and Eldridge, 2009).

In the late 1980s UK cities faced many economic and social challenges due to the effects of de-industrialisation and the 'hollowing out' of cities due to the exodus of people to the suburbs and to new housing estates (Kreitzman, 1999; Roberts and Eldridge, 2009). A new vision, inspired by the journalist and urban campaigner Jane Jacobs, signalled 'a significant move from nine-to-five industrial labour and weekend leisure to a more fluid context of work articulating with 24-hour city leisure practices' (Rowe and Bavinton, 2011: 812). The NTE formed part of the 24-hour city vision, and it aimed to mirror successful European cities at night that had thriving, family-friendly, creative zones in their city centres (Hobbs et al., 2005a). An influential report by the Comedia consultancy (Comedia, 1991) mapped out the complex changes needed, from structural changes such as improved lighting to new policies and legislation, including the relaxing of planning and licensing regulations that were designed to enable a competitive market to flourish (Roberts, 2009). But the free market allowed 'alcohol-centred businesses [to consolidate] their pre-existing dominance, favouring spatiotemporal agglomeration and the conversion or expulsion of alternative uses' (Hadfield, 2015: 1). Some criminologists blamed planners for encouraging 'a reckless growth in alcohol-related entertainment, with dire consequences for crime and disorder'. But, Marion Roberts, professor of urban design and policy, defended planners citing their lack of experience in dealing with entertainment activities within public space and their disempowerment by government in influencing legislation (Roberts, 2009: 488–493). While the causes are complex the outcome was clear, the free market had resulted not only in dominance of alcohol-centred business in some city centres, it had also reduced the diversity of independent venues and small operators for groups such as the ethnic music scene and other creative ventures (Talbot, 2006). By the late 1990s, crime and alcohol-related disorder in the NTE was drawing considerable government, media and public attention, with cities such as Manchester reporting rapidly escalating crime rates (Davies and Mummery, 2006). Urban governance partnership in the NTE resulted in the proliferation of FBO engagement, with Street Pastors formed in 2003 and Street Angels in 2005. Since that time many smaller, local faith-based initiatives including night churches, nightclub chaplains and student-led projects have emerged, but less is known of their work as they are significantly under-researched.

Historically, NTE research in the UK has concentrated upon investigating security, alcohol consumption and policing issues (arguably both secular and male interests). Recent research is now more socio-spatial in perspective and increasingly focuses upon 'the complexity, messiness and local specificity of the formation of urban nights' with key themes of power relations, social exclusion, social sustainability and gender (van Liempt et al., 2015: 409). People groups largely ignored by previous research, such as the homeless, young people and ethnic and religious minorities, have been given new attention (Urban Studies, 2015). There has also been a greater focus on women at work and at leisure in the NTE, including sex workers (Hubbard and Colosi, 2015), door staff (Hobbs et al., 2007) and party goers (Roberts, 2015). Not only has religion been ignored in terms of the social and supportive role that it can provide, but also the fact that many churches hold their prayer meetings, Bible studies and some services in the evening too and are thus their congregants are part of the city in the evenings too, is also ignored. Many older women are participants in these activities and are thus 'out at night' too (cf. Chapter 6).

Networked urban governance

The term 'networked urban governance' (NUG) refers to the formal and informal relational networks and partnerships between public, private and community organisations that are primary vehicles for urban policy development and implementation (Collinge and Hall, 2018). Some prefer the term 'collaborative urban governance' or 'inclusive city leadership' (Ansell and Gash, 2008). But whatever term is used, there are fundamental issues to consider regarding the relative levels of power of the different groups and organisations involved, not least the issue of gender, and the male leadership of established decision-making groups, as against, the more likely female leadership of newer community based groups.

Networked urban governance is a term associated with 'Third Way' governance introduced under New Labour in the mid-1990s, and contributed to neoliberal policies being implemented that resulted in a 'profound reshaping and reweaving of relations between market, state, civil society and individuals' (Gauthier et al., 2013: 15). While neoliberalism can be variously understood as an economic theory, a political ideology, a policy paradigm and a social imaginary (Evans and Sewell Jr, 2013), I was interested in how it has shaped networked urban governance, promoted urban partnership working and increased visibility of faith in the public sphere. Local government actors increasingly adopted the role of 'risk takers and creative inventors' (Hall and Hubbard, 1998: 2) with formal and informal partnerships between public, private and voluntary agencies being central to achieving the goal of competitive, global cities based on consumption and leisure (Hudson and Lowe, 2004: 133). This vision included the goal of a vibrant night and evening economy as part of the strategy to regenerate UK cities (Hobbs et al., 2005b). Marion Roberts attributes the emergence of the UK NTE to both the neoliberal agenda and also the 'cultural turn' towards a 24-hour city with the ideals of a cultural quarter and a creative city (Roberts, 2009).

But, NUG is a highly contested concept, some proponents arguing that it brings about new collaborative working patterns and forms of leadership, and particularly encourages a more entrepreneurial approach to the provision of goods and services. But others see NUG as undermining local government powers and thus democracy. The benefit of urban partnerships and networks is contested, they may be seen to promote change and provide new opportunities for a variety of urban actors but they are also criticised for creating fragmentation and undemocratic systems of control (Deas and Headlam, 2015). Networked urban governance can be researched from different perspectives, and this chapter adopts a network perspective that focuses on my faith practice and my relational network (Pierre and Peters, 2000).

My study was also informed by assemblage thinking (McFarlane, 2011), which provides an 'alternative descriptive lens' which focuses upon the fluidity and complexity of changing urban networks and power groups. I was also interested in Foucauldian understanding of power relationships as applied to the networks and urban phenomena that I was observing (Foucault, 1980). I was finding a complex assemblage of factors and networks, of 'structural shifts and towing undercurrents' that were emerging in respect of the converging (or repelling) of religious beliefs and secular, neoliberal urban governance values (Gauthier et al., 2013: 3). Nationally, there is a 'persisting paradox': 'on the one hand the process of secularization continues; on the other, religion persists as a topic of discussion, indeed dispute, in the public sphere' (Davie, 2015: xiii).

Faith in the public sphere

'Any attempt to understand faith in the public space of civil society and public policy [necessitates an understanding that] … faiths are diverse' (Dinham, 2009: 17) and definitions of faiths, religions, spiritualties are complex. From an urban governance perspective 'faith' has come to represent the main religious traditions recognised by government (Dinham, 2017). Urban governance partnerships have seen an increased number of faith-based organisations (FBOs) in the public sphere since the 1990s when the government showed new interest in faith groups as civil society participants in urban governance (Chapman and Lowndes, 2008). FBOs are variously defined, but for the purposes of my case study, I understood them to be 'any organisation that refers directly or indirectly to religion or religious values, and that functions as a welfare provider and/or as a political actor' (Beaumont, 2008: 2020). This re-emergence of faith in the public sphere is contested with concerns including: their lack of accountability, proselytising, lack of equality and inclusion, and hidden agenda (Chapman, 2009; Farnell, 2009). The place of FBOs in the NTE raises concerns that resonate back to the 'moral street sweepers' of the nineteenth century (Johns et al., 2009) and frankly to concepts of 'sin'. An alternative perspective argues that faith groups make a significant and positive contribution to civil society and that their presence in the public sphere opens up 'the possibility of the enrichment of mutual learning – with new 'knowledges' including religious perspectives, being explored' (Furbey, 2009: 35). Working as a 'faith actor' in the public sphere, I was subject to both social and spatial factors. Space is 'the product of the intricacies and complexities of social relations … relations which are active practices, material and embedded' (Massey, 1999: 283). Previously Massey (1984: 16) defined this process as 'the reproduction over urban space of social relations' with particular reference to gender and power relationships between different groups. Relational networks are understood to be active sites of unstable and contested power relations through which the production of social identities are constantly (re)negotiated (Valentine, 2013).

A *tangled web*

Insights from this summary reveal a complex relational web that significantly enabled and constrained my faith practice. The NTE policy context and the related NUG partnerships supported and legitimated my setting up of the Night Church as well as increasing my agency as a faith leader. But partnership working is also seen to be complex, manifesting entangled relationships and differing values among the groups involved. However, it was my relationships with local faith institutions that were most contested and problematic, due to their pursuit of parochial self-interests and gendered power relations related to my identity as *woman* faith leader. In adopting a feminist perspective to explore my lived experiences as a woman, it is recognised that the terms 'women' and 'women's experiences' are complex and contested. Reflecting on gender from one's own experience is also complex, and such autobiographical accounts 'require a high degree of reflexivity if they are to bring fresh insights' as well as acknowledgment that there is no 'unmediated innocence' to the self and to experience (Fisk, 2014: 9). But the strength of reflecting on 'the living human web' is its capacity to uncover new narratives (Graham, 2014: 197–199) and to use these to review and inform practice.

The Night Church example

I pioneered 'Night Church' in 2010, as an inclusive, Christian sacred space in the heart of the city, based in an ancient Anglican church, starting with two Saturday nights a month between 9.30 pm and 2 am. The vision for Night

Church was to offer an informal space of welcome, safety and hospitality. To facilitate this, it had a café zone with free hot drinks, a creative arts zone, a prayer zone with candles and a quiet seating area. Night Church began as an experimental project, to see whether a welcoming sacred space might be something that people out and about in my own city at night would engage with. Night Church quickly became established in the NTE with between 80 and 200 visitors a night from people out and about in the city. Visitors varied every evening and were from a broad age range but with many in their twenties and thirties, in keeping with the main demographic of the local NTE. People stated different reasons for coming into Night Church including that they were simply curious, thought it was a bar, needed a quiet space, were upset and in need of support. Sometimes individuals came for help with personal problems (such as bereavement), other times entire groups arrived (such as a group of teachers out for the night together), and other agencies such as the police brought people in who they thought the church could help (such as a young man out of prison on licence and unlikely to make it back in time). Only a few people visited regularly, and these were mainly people who were living in hostels, rough sleeping or socially isolated.

My previous work as a director of a local Christian charity meant I had existing links with the local Christian faith network and to some extent with the local authority through its funding of youth and community projects that the charity ran. I was the first woman to be a director of the charity and over the years of working in it I had increasingly noticed attitudes and behaviours that seemed to marginalise women's voices and a culture that was male-centred. The charity was closely associated to a (predominantly male) evangelical faith leaders' network that was very active in the city and had strong links with the council. At the time of pioneering Night Church, I was a member of an Anglican church in the city but had no formal leadership position. Part of my inspiration for Night Church was to create a space that was significantly more inclusive than many of the local expressions of Christian faith, and so I chose its tagline to be '*Night Church is an inclusive Christian sacred space; welcoming to people of all faiths and none*'. Four years later the leadership team expanded the values to: embracing a spiritual life; creating safe space; simplicity; commitment to inclusion; living with questions; dedicated to transformation; working for peace; and celebrating unity in diversity.

In 2014, I resigned as director of the larger charity and led Night Church as an independent faith practitioner in the form of a small, independent charity. This change of positionality and a lack of any clear faith affiliation were to prove significant in the contested relations that emerged. Night Church has always been volunteer-led, initially made up of students and staff from the larger charity along with some other contacts from local churches. Gradually the volunteer team broadened out to include people from a wide range of local churches and some people who had visited Night Church on a night out and asked to become part of the volunteer team. Night Church was pioneered without any specific funding and so I relied on support from others including local shops for food. As the project became established our need for financial support grew and was met from a variety of sources; up to a third came from donations from people who visited on a night (even though it was free and we did not ask for donations, many people asked to give), individuals locally who knew about the work, several local churches who were linked through our volunteers and occasionally small grants from a variety of sources including the police commissioner, the local council and other local trust funds.

Analysing my relational networks in faith practice

I now analyse three relationship networks that were key in my faith practice, that either supported, frustrated or at least impinged upon the running of my Night Church, namely the NTE network, the local urban governance network and the local faith network. In practice these networks are deeply interrelated, but the three-way focus aims to provide depth, comparisons and new insights.

The ambience of the night time economy

In pioneering Night Church, I had not anticipated the effects of its alternative time-space with less hurry, resulting in a greater sense of sociability and solidarity among strangers. The relationships in Night Church appeared to be entirely different from a daytime church café space; people talked to strangers, engaged in lengthy dialogue on many issues including politics and religion and were generally more open to engage in conversation. The sacred space also appeared to provide people with a space to express grief, with many lighting candles as an expression of loss. Having a team of volunteers from a wide variety of backgrounds including doctors, nurses, teachers and trained door staff meant that there was a pool of knowledge to draw on and this helped us to signpost people to appropriate services they might need. My own background as a health service professional helped me to oversee the work of Night Church and deal with issues such as young people who had lost contact with their friends or had fallen out with them and had no means of getting home. Collaborative working with the police, Street Pastors, local clubs and door staff helped to resolve many problems.

A tension for Night Church was retaining its identity as a community space that worked collaboratively with others while avoiding an identity as a specific service provider. Street Pastors were keen for Night Church to develop into a formal 'safe space' with a professionally led triage team as a way of reducing hospital admissions. It was a good vision, but as a team we felt it would move Night Church too much into a service provider model with too much loss of its community identity. Night Church provides a quiet and peaceful social space for those needing a short rest, including those working in the NTE such as door and bar staff, street entertainers and the police. One of the identified problems with the NTE is the lack of diversity of venues and so retaining Night Church as an alternative community space, distinctive from commercial businesses and professional governmental services, was seen to be a priority and a distinctive value of Night Church.

The relational network in urban governance

My relational networks with (secular) urban governance bodies provided important support to set up and sustain Night Church. The Street Pastor team leader facilitated a meeting with the council, and the council's existing relationship with Street Pastors appeared to enable quick support of Night Church. Once this was established, links with others including the police, the homelessness service, and the drug and alcohol treatment services easily followed. The police offered advice on security and provided a city-link radio free of charge. There was already a Street Pastors organisation running in the city and, for example, they helped fund our T-shirts and provided training for volunteers.

My experience of governance partnership working was not always straightforward and collaborative; there were often signs of fragmentation most noticeably on the issue of homelessness. As the leader of Night Church, I was invited to be a trustee for a local homelessness service which led to the council asking me to join a homelessness partnership board. Rough sleepers were regular visitors to Night Church, and I had a good understanding about local faith group involvement in homelessness through my work in the larger faith-based charity. The partnership board spent much of its time grappling with the competing agendas of those involved with rough sleepers in the city, and there was considerable concern about faith groups who were seen to be 'promoting' homelessness by providing food and shelter. The Business Improvement Districts (BID) group was very outspoken in their desire to remove all rough sleepers from the city without any consideration for where they would go and there seemed to be considerable pressure on the police to solve the complex problems. Being on the board helped me to understand why years earlier when Night Church opened, someone in the council sent me a detailed policy on homelessness for us to follow including keeping a log of names of all rough sleepers, and I refused to adopt the policy. This was illustrative of a wider problem, the lack of relationship and communication between the local council and local faith groups regarding homelessness, with deep distrust on both sides.

In network urban governance research, some suggest there is a continuing role for local government in defining priorities and mediating between plural groups (Jewson and MacGregor, 2018; Pierre and Peters, 2000). I am inclined to agree with this given the danger of fragmentation and the complex nature of partnership working, but my experience suggests that a lack of understanding of faith groups and their values significantly hampers local government leaders in this task. Jewson and MacGregor state that 'what is essential to partnerships is knitting together of partnership members in the interests of effectiveness and integration … [to build] coalitions of interests or alliances, among these groups in the name of the wider interest of the city as a whole' (Jewson and

MacGregor, 2018: 9). For this to be achieved local government leaders will need to give much greater attention to building working relationships, 'developing a deeper sense of trust and rapport between local government and religious groups … equipping policy-makers with better understanding of religious groups' (Jawad, 2012: 561).

This understanding may be facilitated by attending to the beliefs, values and world views expressed by a range of faith actors (Baker and Power, 2018) rather than treating faith groups as one homogenised group. This lesson is important to improve local government officers' understanding and decision making regarding 'religion' enabling a more inclusive and diverse faith representation at partnership level. Analysis of the faith network context revealed how difficult it was for some faith actors to deal with other faith institutions, when they themselves had limited power and size, or to enable them to express agency and thus influence decisions. Chapman and Lowndes (2008) suggests that networked urban governance may favour larger and more powerful faith groups while smaller faiths and newer faiths have a harder time being recognised and included. More specifically they suggest that the voices of young people and women are particularly lacking in faith representation and that there was a tendency of policy makers to gravitate towards leaders who were articulate and well networked within the city elites. These findings are supported by this study.

The nature of the faith community network

My most contested relationships emerged from my own local faith network. While there was support from my volunteer team and other FBOs such as Street Pastors, my relationship with two local faith institutions and their leaders was marked by tensions, and being a woman did not help, as discussed below.

Faith institutions

Two very different faith institutions were seen to be a constraining influence on my faith practice. Firstly, the local Anglican diocese, associated with Night Church because of the project being situated in an Anglican building and my affiliation to another Anglican church in the city. When I left the larger charity, I approached the diocese seeking to forge a relational link with them, but this was unsuccessful because they said they did not feel they had the capacity for such a partnership. About a year later, and five years after Night Church opened, the diocese put a church plant (a church mission group from elsewhere without their own building) and an ordained pioneer minister into the same church building as was being used by the Night Church. I only heard of the change on the grapevine a few weeks before it happened. This change proved significant because the pioneer leader had his own vision and showed no interest in working collaboratively with Night Church. As a result, the creative space for Night Church became considerably restricted and so a year after this change I handed over the leadership of Night Church to members of my team in order to pursue my ordination studies.

The second faith institution was the evangelical faith leaders' network who described themselves as a 'one-stop-shop' for faith in the city. Its leadership was male and in the network's churches women were very significantly under-represented or absent in senior leadership roles. In my previous work I had found this group challenging due to their marginalisation of women as leaders and my relationship with the leaders was strained. When I pioneered Night Church this faith leaders' network showed interest in the project and expressed concern that it was called 'church'. Once Night Church became an independent charity, they questioned its accountability structures and suggested that it should 'come under' the authority of a church in the network. There seemed to be an assumption that all FBOs were affiliated and somehow accountable to them, illustrated most clearly when they held a meeting for all FBO leaders in the city. They suggested that there were now too many of us 'fishing in the pond' and that we were taking too many of their church members as volunteers. It was suggested that some of us should perhaps close. Thus gender played a major part in the chances of my own faith practice being taken seriously.

My experience suggests that the relationship between faith institutions and more entrepreneurial faith actors working on the margins of those institutions may produce considerable power shifts and contested power relations. A tendency for faith institutions to defend their own narrow parochial interests is problematic given that in a UK context 'the forms of religion which are growing fastest ... allow considerable lay participation, blur the lay–clerical boundary, are entrepreneurial, experiment with new, often charismatic, forms of authority, and are often rather institutionally-fragile' (Woodhead, 2016: 45).

While I self-identify as a faith practitioner, with an emphasis on praxis (that is putting faith into action), I also relate closely to the concept of the faith entrepreneur (and more broadly the social entrepreneur). Research on entrepreneurship has begun to pay increased attention to the role of religion and spirituality (Henley, 2017) and how it shapes entrepreneurial motivations, mindsets, attitudes and behaviours and values and beliefs (Balog et al., 2014). The internal values of religion are seen as primary drivers for entrepreneurial engagement, which I relate to in my own faith engagement in Night Church which was motivated by my own faith and my desire to make a sacred space more accessible to the community. Balog et al. (2014) highlight that entrepreneurs place a deep level of personal meaning on their pursuits and this can be seen in the commitment of myself and also many other members of the team who were willing to work as volunteers from Saturday evening through to 3 am on a Sunday morning.

Gendered power relations

In the contested power relations with the local faith network there is a 'towing undercurrent' of gender, which held me back, and was the result of my identity as a woman and my role as a faith leader. This conclusion draws on my understanding of gender as being socially constructed through repeated acts and performances 'that congeal over time to produce the appearance of substance, of a natural sort of being' (Butler, 1990: 33). Exploring my experience as a *woman* faith practitioner from a feminist perspective brings a commitment to listen to the lived experiences of women's lives and to bring 'silence into speech' in order to challenge dominant cultures that may render women's lives 'unspeakable' (Graham, 2014). But my lived experience is understood to be particular and local, acknowledging that understandings of the term 'women' and 'women's experiences' are complex and diverse.

Conclusion

My experience of faith practice across the three parts of my relational network reveals a marked contrast in my experience of faith leadership. In the NTE and in governance relationships I experienced increased agency and opportunities as a faith leader, and an environment where it was easy to build partnership working. In contrast, with faith institutions it was difficult to form partnerships and I sensed an 'invisibility' and a resulting lack of agency partly because of my gender and my role as an independent 'outsider'. My experience supports the idea that neoliberalism can be empowering because it 'may help the marginalized groups break free from pre-existing structures of social powers and open new avenues of participation and influence' (Kashwan et al., 2018: 5). However, my insights also suggest that in these new relational spaces the influence of pre-existing power relations may continue to exert considerable power. When viewed from a broad perspective, there would appear to be an underlying gender bias by the faith institutions,

The analysis of my relational network reveals a complex map of relationships circulating around my faith practice, with enabling and supportive new alliances in networked governance and the NTE emerging alongside constraining power relations with local faith institutions. Some key insights emerge from the case study that may merit further research. While the case study was particular and local, it provides new insights into networked urban governance and partnership working from a faith practitioner's perspective. The NTE provided me with a creative space and positive partnerships that enabled my entrepreneurial faith practice. The case study illustrates how faith in the public sphere may widen participation and agency for some faith actors including women who find it difficult to appropriate faith leadership in their own faith structures. However, such faith actors may need the skills to negotiate complex and contested power relations and to build strong alliances in governance partnerships, within a medium-sized city within the northwest of England

Networked urban governance partnerships and my relational network in the NTE provided stability and sustainability for Night Church and increased agency and recognition for me as a woman faith practitioner. But the local council's lack of understanding of faith in its local context and its over-dependence on the established 'evangelical faith leaders' network limits its capacity to build coalitions of interests across a more diverse faith network. An engagement with faith groups beyond an instrumental discourse and giving greater attention to beliefs, values and worldviews would arguably enhance communication and build trust. Local governance leaders need to be informed enough about faith groups to engage critically with faith representatives in regard to 'which interests and which players will be included in partnerships and which will be left out?' (Jewson and MacGregor, 2018: 9).

My experience suggests that faith practitioners operating on the margins of faith institutions may be particularly vulnerable to complex power relations, especially if they are women. Parochial, faith institutions acting in self-interest may be protective of resources including buildings, leadership structures, membership resources and relational networks. In particular, faith practitioners working in FBOs may be viewed as a competitive threat to those spatial and relational resources, with their new faith practice producing significant power shifts and movement of resources. This highlights the complexities of faith in the public sphere with entangled relationships and competing beliefs, values and worldviews being enacted by faith actors from a diverse faith network. Some women faith practitioners may face additional challenges due to contested power relations not only related to more innovative faith practices but also due to gendered power relations.

References

Ansell, C. and Gash A. (2008) 'Collaborative governance in theory and practice', *Journal of Public Administration Research and Theory* (18): 543–571.

Baker, C. and Power, M. (2018) 'Ghosts in the machine: beliefs, values and worldviews in the workplace and business environment', *Journal of Beliefs and Values* 39(4): 474–489. doi: 10.1080/13617672.2018.1477705.

Balog, A., Baker, L. and Walker, A. (2014) 'Religiosity and spirituality in entrepreneurship: a review and research agenda', *Journal of Management, Spirituality and Religion* 11(2): 159–186.

Beaumont, J. (2008) 'Faith action on urban social issues', *Urban Studies* 45(10): 2019–2034. doi: 10.1177/0042098008094871.

Butler, J. (1990) *Gender Trouble: Feminism and the Subversion of Identity*, London: Routledge.

Chapman, R. (2009) 'Faith and the voluntary sector in urban governance: distinct yet similar' in A. Dinham, R. Furbey and V. Lowndes (eds) *Faith in the Public Realm: Controversies, Policies and Practices*, Bristol: Policy Press, 203–233.

Chapman, R. and Lowndes, V. (2008) 'Faith in governance? The potential and pitfalls of involving faith groups in urban governance', *Planning Practice and Research* 23(1): 57–75. doi: 10.1080/02697450802076480.

Collinge, C. and Hall, S. (2018) 'Hegemony and regime in urban governance: towards a theory of the locally networked state' in N. Jewson and S. MacGregor (eds) *Transforming Cities: Contested Governance and New Spatial Divisions*, Abingdon: Routledge, 129–140.

Comedia (1991) *Out of Hours: A Report for the Gulbenkian Foundation*, London: Demos and Comedia.

Davie, G. (2015) *Religion in Modern Britain: A Persistent Paradox*, Chichester: Wiley Blackwell (2nd edition).

Davies, P. and Mummery, H. (2006) *Night Vision: Town Centres for All*, London: The Civic Trust.

Deas, I. and Headlam, N. (2015) 'Boosterism, brokerage and uneasy bedfellows: networked urban governance and the emergence of post-political orthodoxy' in R. Paddison and T. Hutton (eds) *Cities and Economic Change: Restructuring and Dislocation in the Global Metropolis*, London: Sage, 127–144.

Dinham, A. (2009) *Faiths, Public Policy and Civil Society: Problems, Policies, Controversies*, New York: Palgrave Macmillan.

Dinham, A. (2017) 'Re-evaluating value in faith-based social action', *Voluntary Sector Review* 8(2): 205–218. doi: 10.1332/204080517x14972755356351.

Dinham, A. and Lowndes, V. (2009) 'Faith and the public realm' in A. Dinham, R. Furbey and V. Lowndes (eds) *Faith in the Public Realm: Controversies, Policies and Practices*, Bristol: Policy Press, 1–20.

Evans, P. and Sewell Jr, W. (2013) 'Neoliberalism: policy regimes, international regimes and social effects' in P. Hall and M. Lamont (eds) *Social Resilience in the Neoliberal Era*, Cambridge: Cambridge University Press, 35–68.

Farnell, R. (2009) 'Faith, government and regeneration: a contested discourse' in A. Dinham, R. Furbey and V. Lowndes (eds) *Faith in the Public Realm: Controversies, Policies and Practices*, Bristol: Policy Press, 183–203.

Fisk, A. (2014) *Sex, Sin and Our Selves: Encounters in Feminist Theology and Contemporary Women's Literature*, Eugene, OR: Pickwick Publications.

Foucault, M. (1980) Two lectures (C. Gordon, L. Marshall, J. Mepham, and K. Soper, Trans.) in C. Gordon (ed.) *Power/Knowledge: Selected Interviews and Other Writings 1972–1977*, New York: Pantheon, 78–108.

Furbey, R. (2009) 'Controversies in "public faith"' in A. Dinham, R. Furbey and V. Lowndes (eds) *Faith in the Public Realm: Controversies, Policies and Practices*, Bristol: Policy Press, 21–40.

Gauthier, F., Martikainen, T. and Woodhead, L. (2013) 'Religion in market society' in F. Gauthier and T. Martikainen (eds) *Religion in the Neoliberal Age: Political Economy and Modes of Governance*, Farnham: Ashgate, 1–20.

Graham, E. (2014) 'Feminist theory' in B.J. Miller-McLemore (ed.) *The Wiley-Blackwell Companion to Practical Theology*, Oxford: Wiley-Blackwell, 193–203.

Hadfield, P. (2015) 'The night-time city. Four modes of exclusion: reflections on the *Urban Studies* special collection', *Urban Studies* 52(3): 1–11. doi: 10.1177/0042098014552934.

Hall, T. and Hubbard, P. (1998) 'The entrepreneurial city and the "new urban politics"' in T. Hall and P. Hubbard (eds) *The Entrepreneurial City: Geographies of Politics, Regime, and Representation*, Chichester: Wiley, 1–26.

Henley, A. (2017) 'Does religion influence entrepreneurial behaviour?', *International Small Business Journal* 35(5): 597–617.

Hobbs, D., Hadfield, P., Lister, S. and Winlow, S. (2005a) 'Violence and control in the night time economy', *European Journal of Crime, Criminal Law and Criminal Justice* 13(1): 89–102. doi: 10.1177/1748895808092428.

Hobbs, D., Winlow, S., Hadfield, P. and Lister, S. (2005b) 'Violent hypocrisy: governance and the night-time economy', *European Journal of Criminology* 2(2): 161–183. doi: 10.1177/1477370805050864.

Hobbs, D., O'Brien, K. and Westmarland, L. (2007) 'Connecting the gendered door: women, violence and doorwork', *British Journal of Sociology* 58(1): 21–38. doi: 10.1111/j.1468-4446.2007.00137.x.

Hubbard, P. and Colosi, R. (2015) 'Taking back the night? Gender and the contestation of sexual entertainment in England and Wales', *Urban Studies* 52(3): 589–605. doi: 10.1177/0042098013504006.

Hudson, J. and Lowe, S. (2004) *Understanding the Policy Process: Analysing Welfare Policy and Practice*, Bristol: Policy Press (2nd edition).

Jawad, R. (2012) 'Religion, social welfare and social policy in the UK: historical, theoretical and policy perspectives', *Social Policy and Society* 11(04): 553–564. doi: 10.1017/s1474746412000309.

Jewson, N. and MacGregor, S. (2018) 'Transforming cities: social exclusion and the reinvention of partnership' in N. Jewson and S. MacGregor (eds) *Transforming Cities: Contested Governance and New Spatial Divisions*, Abingdon: Routledge, 1–18.

Johns, N., Squire, G. and Barton, A. (2009) 'Street Pastors: from crime prevention to remoralisation', *British Society of Criminology* 9: 39–56.

Kashwan, P., MacLeanb, L. and García-López, G. (2018) 'Rethinking power and institutions in the shadows of neoliberalism', *World Development*. doi: doi.org/10.1016/j.worlddev.2018.05.026.

Kreitzman, L. (1999) *The 24 Hour Society*, London: Profile Books.

Massey, D. (1984) *Spatial Divisions of Labour: Social Structures and the Geography of Production*, London: Macmillan.

Massey, D. (1999) 'Spaces of politics' in D. Massey, J. Allen and P. Sarre (eds) *Human Geography Today*, Cambridge: Polity Press, 279–294.

Maynard, M. and Purvis, J. (eds) (1994) *Researching Women's Lives from a Feminist Perspective*, London: Taylor & Francis.

McFarlane, C. (2011) 'Assemblage and critical urbanism', *City* 15(2): 204–224. doi: 10.1080/13604813.2011.568715.

Melbin, M. (1978) 'Night as frontier', *American Sociological Review* 43(1): 3–22.

Pierre, J. and Peters, G. (2000). *Governance, Politics and the State*, Basingstoke: Palgrave Macmillan.

Roberts, M. (2009) 'Planning, urban design and the night-time city: still at the margins?', *Criminology and Criminal Justice* 9(4): 487–506. doi: 10.1177/1748895809343415.

Roberts, M. (2015) '"A big night out": young people's drinking, social practice and spatial experience in the "liminoid" zones of English night-time cities', *Urban Studies* 52(3): 571–588. doi: 10.1177/0042098013504005.

Roberts, M. and Eldridge, A. (2009) *Planning the Night-time City*, Abingdon: Routledge.

Rowe, D. and Bavinton, N. (2011) 'Tender for the night: after-dark cultural complexities in the night-time economy', *Continuum* 25(6): 811–825. doi: 10.1080/10304312.2011.617875.

Schlor, J. (1988) *Nights in the Big City*, London: Reaktion.

Talbot, D. (2006) 'The Licensing Act 2003 and the problematization of the night-time economy: planning, licensing and subcultural closure in the UK', *International Journal of Urban and Regional Research* 30(1): 159–171. doi: 10.1111/j.1468-2427.2006.00642.x.

Urban Studies (2015) *Special Issue: Geographies of the Urban Night*, Vol. 52.

Valentine, G. (2013) *Social Geographies: Space and Society*, Abingdon: Routledge.
van Liempt, I., van Aalst, I. and Schwanen, T. (2015) 'Introduction: geographies of the urban night', *Urban Studies* 52(3): 407–421. doi: 10.1177/0042098014552933.
Woodhead, L. (2016) 'Intensified religious pluralism and de-differentiation: the British example', *Society* 53: 41–46. doi: 10.1007/s12115-015-9984-1.

14 Faith, gender and the planning system in the UK

Carey Doyle

Abstract

Social planning in the United Kingdom (UK) has remained peripheral to planning practice, despite new challenges from increasingly complex diversity. For faith planning specifically, there is divergence between diverse religious needs and the institutionalised and secular spatial logic of planning. This chapter investigates faith, gender and planning practice in three UK case studies with substantial population change and diversity. In the case studies planners relied on 'community leaders' representing groups as a key means of engaging with diversity, with attendant conflict shown in the literature to be common with the use of community leaders. These practices were further challenged by diversity; the increased numbers, varying natures and spatio-temporal differences of groups presented challenges to engagement practices premised on simplistically conceived social groups represented by leaders. Moreover, a notable amount of community leaders were female, and unequal outcomes for these leaders underline the possible negative implications for females engaged in civil society in this way. Where minority ethnicity, faith and female gender intersect these issues can be pronounced. Initial recommendations for further research and practice are provided in conclusion.

Key words: planning practice, immigrants, diversity, community leaders, planning law, consultation

Introduction

The UK is increasingly socially and culturally diverse, leading to new challenges in the provision of public services due to varying nature and number of needs (Vertovec, 2007). The ability of the devolved planning systems to accommodate these diverse needs is limited by a number of issues. There is a long-established practice of planners considering their work objective to social impacts, despite decades of evidence that in practice that planning acts unequally on social groups. Where planning does engage with social issues, the impacts can be uncertain and often negative, despite positive intentions (Greed, 1994; Thomas, 2000;

Gale, 2008; Beebeejaun, 2012; Fainstein, 2000; Thomas and Gale, 2015; Gale, 2005; Gaffikin et al., 2016). The heterogeneity of contemporary diversity further challenges social planning practice, and heightens the implications of failures (Doyle, 2018). Where planning fails to grapple meaningfully with diversity, it is a missed opportunity to use the tools of planning to create better, more inclusive, communities.

These issues are notable for faith planning research, which has provided important insights into how religious spatial needs intersect with the institutionalised spatial logic and implicit secularity of planning in the UK (Gale and Naylor, 2002; Gale, 2008; Dwyer et al., 2013; Greed, 2016a). This chapter furthers these insights by investigating faith, gender and planning practice in three case study locations with substantial population change in England and Northern Ireland. Drawing on intersectional theory, the chapter focuses on meaningful intersections of social heterogeneities; it applies a lens to intersections of faith and gender but also notes cogent intersections with other social characteristics (particularly ethnicity), and pays attention to power.

In the case studies planners commonly relied on 'community leaders' to represent perceived social groups (for faith groups, as well as other groups, such as ethnicity or 'migrants'). These simplistic social group understandings applied in planning practice could produce conflict between, and exert an exclusionary force on, the multiplicity of faith groups. Complexity in the number, nature and spatio-temporal characteristics of groups presented further challenges to planning practice. Moreover, there were notable female community leaders, with unequal outcomes for these leaders, including unpaid labour and disenfranchisement. This paper argues that there can be amplified limitations to planning practices in places of population change and increased social diversity, which can be compounded for faith groups and females.

This chapter firstly reviews literature on faith and planning practice in the UK, as well as on engagement and the use of community leaders for engagement. Methodology is then provided, including detailing the case studies. The challenges of using community leaders in contexts of social heterogeneity in the case studies are then outlined. This is followed by a discussion of the implications of community leadership for female participants. The chapter concludes with a discussion, and identifies areas for future work.

Faith and planning practice in the UK

The devolved nations of the UK (England, Wales, Scotland and Northern Ireland) have their own planning systems. These are all implicitly secular, with land-use planning operating as an administrative system controlling the development of land. Planning functions officially in the public good (e.g. 'to achieve sustainable development'), but in practice it operates in the interests of those who can exert the wealth and power to produce results (Fox-Rogers and Murphy, 2014). Planning in the UK has a widespread perceived objectivity to social issues (Greed, 1994; Thomas, 2000; Ellis, 2015; Thomas and Gale, 2015;

Gale, 2005; Gaffikin et al., 2016). For faith specifically, this is a false secularity in the implementation of a rational and neoliberal model (McClymont, 2015; Greed, 2016b; Manouchehrifar, 2018).

UK planning is secular to the extent that land-use classifications do not address religious use specifically. The case studies are in two different devolved administrations (England and Northern Ireland). In England faith facilities including places of worship and supporting faith spaces are grouped with other 'non residential institutions' (use class 'D1'), which includes clinics, health centres, crèches, day nurseries, day centres, schools, art galleries, museums, libraries, halls, places of worship, law courts, non-residential education and training centres. The Northern Irish planning system is widely considered by practitioners to be separate to the entrenched ethno-national and religious divisions in that jurisdiction; here secularity can be perceived to be necessary for fairness (Gaffikin and Morrissey, 2011).

At the end of the twentieth century legislation to address unequal impacts to social groups was introduced in the UK, including faith and gender specifically. In England the Equality Act 2010 established a requirement for public authorities to have due regard to eliminating discrimination and advancing equality of opportunity, with a series of 'protected characteristics' identified, which includes religion or belief, and sex (meaning gender). In Northern Ireland relevant legislation includes Section 75 of the Northern Ireland Act 1998 and the Race Relations Order 1997. The Northern Ireland Act 1998 requires public authorities to have due regard to the need to promote equality of opportunity between a number of groups, including those of different religious beliefs, and men and women.

In practice considering the social impacts of planning policies and projects frequently relies on 'protected characteristic' groups derived from equality legislation (Gale, 2008; Beebeejaun, 2012). Equality legislation provides an important legal basis, but expertise and guidance in planning practice on how to consider impacts is not widely held (O'Neill and Murtagh, 2017). The protected characteristics can be simplistically conceived, with limited links to evidential data and limited understanding of possible impacts arising from planning policies or projects. The application of UK equality policy has been shown to have a common focus on procedure not substance, within questionable links to outcomes and lack of political will (Radaelli and Meuwese, 2009; Shortall and McAreavey, 2017).

Protected characteristics are often used as the basis for social planning, including for engagement practices that seek to recognise and accommodate diversity by targeting identities or groups. Those with these characteristics can be considered vulnerable, excluded, less likely to engage or 'hard to reach'. In this way social group labels derivative of a single social attribute (such as faith or ethnicity) become identities that are applied, contested and perpetuated in engagement practices. There is a substantial literature that has explored the limitations of a policy focus on a single social category. This has noted that planners often rely on a small number of social groups, typically uncritically considered to be objectively determined (Beebeejaun and Vanderhoven, 2010). Engagement focused on social groups may over-emphasise differences between groups and underplay similarities within groups (Uitermark et al., 2005: 624). There are practical challenges arising from engaging with an ever-increasing number of social groups, such as where specific languages or ethnicities are targeted (Phillimore, 2013). There is a need to accommodate differentiation (Murtagh et al., 2008; Gale, 2008; Beebeejaun and Grimshaw, 2011).

Identifying groups can be as much of an exercise in control and management as in recognition (Brownill and Carpenter, 2007: 630). There are limitations of targeted consultation, questioning the extent to which meaningful recognition occurs and who benefits. With regard to power, it has been argued that consultation is not simply empowering, but that dangers can arise (Sandercock, 2003; Fincher and Iveson, 2008; Baker et al., 2010; Fincher et al., 2014).

There is a need to better consider social identities that do not easily align with a single social category. Using an approach grounded in intersectionality (Cooper, 2015), we can trace intersections of social characteristics and the operation of power. In this chapter gender and faith are the prime focused, but other social characteristics are included as well where these arise in the case studies. These social heterogeneities and intersections thereof provide variable experiences and needs that are difficult to address in singular, objective approaches which assume the presence and nature of an individual association with a particular social category.

Targeted consultation techniques in the UK often use civil society and community leaders to engage with identified groups. Using community leaders is to a degree a matter of practicality, as planners typically do not have (and pragmatically could not have) established working relationships in all neighbourhoods and with all religious, language or cultural communities. However, consultation targeted at social groups assumed to be represented by a 'community leader' has been noted to engage a politics of community leaders (Uitermark et al., 2005; Murtagh et al., 2008; Gaffikin and Morrissey, 2011; Thomas and Gale, 2015). Community leaders can include or exclude individuals and other groups. There is a benefit to stifling internal group heterogeneity, as policy incentivises those who present the most cohesive community. The literature notes that these leaders hold a powerful position, as they can exert a degree of power over the communities they represent.

Methodology

This chapter draws on three case studies. Two case studies were part of a research project on planning, migration and diversity, with fieldwork undertaken in 2014–2015. One of these case studies was in rural England and one in rural Northern Ireland. A third urban English case study was conducted as part of a professional planning project on faith and planning in 2017 (CAG, 2017). All three case studies were locations of substantial population change and diversity; one notable measurement is that all three local government authorities were within the top five highest rates of immigration growth between the 2001 and 2011 Censuses, within England and Northern Ireland.

The research from 2014–2015 included 71 semi-structured interviews in two UK rural locations, one in England and one in Northern Ireland. Thirty-nine interviews were conducted in Northern Ireland (10 regional civil servants, 10 officers and councillors in local government, 5 from non-governmental organisations and 14 migrants from five countries of origin) and 32 in England (1 regional civil servant, 12 officers and councillors in local government, 8 from non-governmental organisations and 11 migrants from seven countries of origin). Consent was granted by all participants. An honorarium of £20 was offered to migrant participants as a token recognition of their time and effort.

The 2017 research involved 27 interviews, a survey of faith groups (63 participants), two workshops with faith groups and a number of meetings with the local authority. This research was conducted on behalf of the local authority. The project was conducted in English. Consent was granted by all participants for involvement in the council-sponsored project. I was the project lead, with direct access to the local authority stakeholders as well as faith group participants. Intellectual property rights for the use of the research for academic purposes have been obtained.

Data for all three projects was analysed and interpreted by reading and re-reading transcripts, field notes and other data, followed by a process of coding to identify and embed emerging issues (Boeije, 2009).

Community leaders, faith and heterogeneity

In all three case studies those in local government and public services (such as planners, elected representatives and employees at non-governmental organisations) frequently used individuals as community leaders to represent faith groups, minority ethnic groups, migrant cultural and language groups and other targeted groups. This reflects the literature that has evidenced the use of community leaders in that UK consultation. In the case studies community leaders were targeted for a wide variety of local authority needs, including for specific policy consultations (such as a new local plan), representation on committees and ongoing informal engagement when issues arose where a perspective was considered to be needed. Examples of informal engagement that occurred are queries by other organisations (such as central government departments or the police), a flooding emergency and interview requests by outside researchers.

Community leaders could be asked to facilitate engagement by others from a group, or to provide a representative perspective for a group. Examples of simplistic social groups used in consultation practices in the case studies were a single faith denomination (e.g. for all Pentecostal Christians), for all minority faith residents in a particular area, or for a language group (e.g. all Polish, including Polish faith needs). In this way policy and engagement strategies exerted a force for community leaders to represent to externally identified social groups.

In the case studies there were a number of challenges to this engagement approach borne out of increased social and cultural heterogeneity. Notably, this heterogeneity comprised not just additional residents, and consequently additional groups to incorporate (such as increased numbers of faith groups). Variation within externally identified groups was increasingly important, such as different languages within a minority faith group, or differences of need for men and women within one faith group. Temporal and spatial factors, such as rates of growth, ongoing transient migration, and geography, further complicated matters. For example pronounced rates of growth meant that local authorities had few established links to new faith groups. These new groups would then move as the group grew or declined. Faith groups could move to new locations within a local area, or outside a local area (in the most extreme cases moving numerous times, transiently). Sometimes groups split into associated groups. Geography had a number of implications, including notable distances between the locations of housing and place of worship, leading to transit costs, emissions and other indirect effects. There were also varying faith congregant spatial scales (e.g. neighbourhood mosques convenient for daily prayer, while a synagogue or Buddhist temple could provide for a regional congregation). These practical implications of increased diversity of faith groups resulted in a number of challenges to planning practice and engagement. Two of these challenges are discussed below: competition and conflict between community leaders and a pressure to not allow for divergent views within groups.

In the urban English case study, a development of 1,000 homes (phase 1 of a substantial new development of 10,000 dwellings), was recently completed. A single multi-faith room had been provided as religious accommodation for this phase, with the intention that all the new residents' faith needs would be accommodated within a shared room. However, there were substantial difficulties with sharing, leading to competition and conflict. There are fundamental issues with sharing worship space, including the requirement of only Christian worship to occur in consecrated Christian worship spaces, and as daily Muslim worship occurs in accordance with set prayer times, it is not possible for different Muslim groups to share space and worship at different times. In this case the shared room was given to a Christian organisation to manage, on the basis of the strong engagement between a particular faith community leader and the local authority. Accommodation was made for some groups to share the space, but not others, with limited transparency on these decisions. There was also lack of clarity on management, as well as when, and how, additional space for faith groups would be provided in the development. This example evidences a failure in planning practice to account for faith group heterogeneity, and instead provide a single multi-faith room. The practical sharing issues point to the importance of community leaders required to broker sharing arrangements, both on a day-to-day basis and to influence governance more strategically. Competition and conflict was notable; but importantly there were also successes in well-established space-sharing relationships between some groups.

The importance of providing for heterogeneity, subjectivity and difference, as well as the role of power in engagement, needs to be accounted for. An extract from an internal meeting note illustrates the cogency of gender, differentiation within faith denominations and power in relations between community leaders:

> At the meeting X attended in her role as head of the Y committee for the council (X was an elected representative, as well as active female leader in a large faith group). She asked for the list of invitees to the project steering committee, and when she received this she became angry saying that invitees should have been approved by her first. One female invitee was of particular concern, a representative of a newer faith group within the same denomination as X, but in a different area. X said she would not attend if this person was on the steering group.
>
> (Project Notes, 2017, Author)

In response to this event, the project team made a decision not to proceed with the project steering group to avoid these types of conflicts. This may have decreased the level of involvement by faith groups in the study, but it also protected attendees from the conflicts that could arise. Not having a steering committee also removed the power dynamics inherent in being requested to be on a committee, or excluded from a committee, by representatives of the local authority. At a later stage in the project a number of faith group participants voiced concern over the structure and content of consultation exercises led by the local authority, noting that sessions which involved listening to faith groups were welcome, but that there was a desire to avoid confrontational meetings as the behest of the local authority.

There were also limitations to the local authorities' engagement arising from the spatio-temporal differences in minority groups. Community leaders and community groups were doorkeepers to wider language and cultural communities, and provided some local services such as childcare and homeless outreach. However, the transience of residents and community groups meant that local authorities did not have a satisfactory understanding of the number and nature of needs in their area. Evidence of this can be taken from a study of council records for 2017, which initially indicated over 150 faith groups in the area, of which 128 were confirmed as continuing to operate, some 22 less than previously identified. Moreover, a number of the confirmed faith groups were in new locations, indicating new facilities, sharing with other groups and other changes that could sever lines of engagement.

Transience presented particular challenges to engagement practices based on leadership and cohesive communities. As noted by a participant:

> It was not like 'traditional' immigration where people would become community leaders or community spokespersons. There was no community as such. Because it was such a transient, dispersed group … some would stay in

> (case study) for three months, some had the intention never to stay in (case study) … But then you had another group that did settle and start to raise a family and put down roots.
>
> (Rural English Case Study, Male, Equality Training Consultant, 2015)

There was an expectation for community leaders representing immigrant communities, but this was undermined by transience.

Social, spatial and temporal heterogeneity presented notable challenges to engagement practices based on community leaders, as well as planning practice which applied simplistic understandings of faith needs. There could be substantive power differentials between those groups with leaders that were successful at engagement and those that were not, with associated conflict.

Female community leaders and the work of being hard to reach

A number of women did substantive community leader work in the divergent case studies. These participants noted variable and often limited recompense for their work as community leaders, raising questions of the gendering and disenfranchisement of community leader labour, particularly for minority faith and ethnic groups.

Engaging with local authorities as a community leader could require substantial effort, as requests for input occurred both formally and informally on an ongoing basis. These efforts at facilitating recognition of needs and perspectives of communities within local government processes were only part of the responsibilities of community leaders, who also undertook other duties such as organising festivals and cultural groups.

The power entailed in requesting engagement from a community leader could be considered inherent in labelling those targeted for consultation as 'hard to reach'. Who is actually hard to reach, and from what perspective? What are people excluded from exactly? One participant put this bluntly:

> Migrants have joined the 'hard to reach' category. The poor are hard to reach … People are not hard to reach if you are standing beside them. So we will always say to council 'Do you know who is hard to reach? The police are very very hard to reach. The government is hard to reach. Social Services are hard to reach. The Council are hard to reach. Gangmaster enforcers are hard to reach'.
>
> (Northern Irish Case Study, Female, Non-Governmental Organisation, 2015)

Consultation fundamentally involves power, and for those faith groups, women and ethnic minorities which engaged, this power was transacted largely by the requesting authority, and on their terms.

Being community leader was variably compensated. Some participants were paid for the role; others had professional roles and volunteered as community leaders. There was evidence of negative implications of being a community leader. For example, a participant described paying out of pocket for a community cultural event (e.g. an Easter party), and never recouped this cost. This participant described this as 'being taken advantage of':

> Lots of the institutions take advantage of me. Usually I do not refuse to take part in local projects and help people. I represent the European communities.
>
> (English Rural Case Study, Female from Lithuania, 2015)

This participant expressed frustration over the difficulty of getting paid employment which reflected her expertise, and how the council and other stakeholders would require her input for free, while employing a range of other consultants. In this way there could be variable outcomes for community leaders, with some benefiting such as through employment, but others were negatively impacted.

Heterogeneity of leaders produced a range of outcomes, with some benefiting, and others impacted. A number of female interview participants described balancing community leadership work with professional work and caregiving demands. Some female participants appeared to do this successfully, reaping professional benefits from being a community leader. Others described frustrations with extracting benefit from the work of being a community leader, and expressed anxiety over negative implications, such as unpaid labour or inability to produce solutions sought in their 'leader' role. Some female participants in the rural case studies emphasised that they were not community leaders, illustrating the complicated associations with these roles. Acting as a community leader could negatively impact on finances, time and personal lives, and there was no guarantee of any outcome from engaging with planning processes.

Discussion and conclusions

Planning practice in the UK includes legal provision for having 'regard' to the elimination of discrimination and advancing equality of opportunity with regard to faith and gender. A key means of incorporating this into planning practice is engagement with diverse groups, often relying on community leaders to provide a link between planners and 'hard to reach' community groups and residents. Policy for engaging has continued to be conceptually rooted in bounded and place-specific social groups, 'cohesive communities'. However, there is substantial evidence in the literature of unequal impacts arising from engagement, as well as limited ability in reducing, and in advertently reinforcing, exclusion. This chapter has outlined some of the ways in which, for faith and gender, these issues have been exacerbated by increased social heterogeneity, further working against the stated aims of equality legislation and planning engagement.

Existing engagement practices perpetuated governance structures of hierarchy according to community leaders and the taxonomy of groups, which community leaders had to use in order to gain power. In the dynamic case studies with substantial population change, there were substantial numbers of groups with varying needs: some community leaders engaged and exerted power, and others engaged but realised few outcomes. Other groups did not have engaged leaders, and consequently were excluded from these processes altogether. There can be substantive power differentials between those groups that are successful at engagement and those which are not – this may occur in most contexts, but where there is increasing social heterogeneity differentials between groups can be pronounced. It is not a matter of a small number of groups jostling for funding, but instead ongoing processes of groups being formed, some dissolving, in different places and at different times, with some groups represented with successful outcomes, other groups not represented at all, and other groups struggling to get legitimation.

Increased attention is needed to how engagement processes can negatively impact community leaders, particularly those who are female and from minority ethnic and faith groups. Those providing representation as community leaders can be personally and professionally impacted by the demands of this work. Community leaders are not only those who benefit from the power allotted to them – some devoted substantial energies with limited recompense other than a personal commitment to helping others. In an era of reduced public services, demands for community leadership can be extensive and consequently have negative impacts on finances, time and personal lives. Planners and local authorities largely dictated these processes.

Identifying unequal impacts, conflicts and exclusions as I have done in this chapter is important in documenting the limitations of current planning practice. However, focusing planning research on conflict and limitations perpetuates an antagonistic understanding of planning. To this end, the chapter ends by considering future research and ways to address these limitations in practice.

It will be valuable in future research to focus on productive planning and means engagement for faith groups and women. This will require applying critical concepts for understanding diversity and planning which are attuned to intersectionality and power. Ultimately planning exists to operate in the public good, and consequently practical knowledge is necessary. There are a number of practical implications of this research. In conducting engagement exercises there is a need to be sensitive to unequal uptake of community leader roles, and the heterogeneity within targeted groups. Community mapping techniques can be used to establish where gaps exist. The work required of community leaders could be better recognised with clearer avenues to outputs. To address power differentials there should be opportunities for involvement outside established community groups and leaders. To counter community leaders' power differentials and to provide a space for those who do not align with simplistic social groups, targeted group consultations can be balanced with cross-group consultations based on commonalities. This provides opportunities for communities to cohere or not cohere, around varying issues. These are initial steps towards a UK planning practice which is attuned to the multi-dimensional ways in which faith, gender and other social characteristics interact with planning and power.

References

Baker, Mark, Stephen Hincks, and Graeme Sherriff (2010) 'Getting involved in plan making: participation and stakeholder involvement in local and regional spatial strategies in England', *Environment and Planning C: Government and Policy* 28(4): 574–594.

Beebeejaun, Yasminah (2012) 'Including the excluded? Changing the understandings of ethnicity in contemporary English planning', *Planning Theory and Practice* 13(4): 529–548.

Beebeejaun, Yasminah, and Lucy Grimshaw (2011) 'Is the "New Deal for Communities" a new deal for equality? Getting women on board in neighbourhood governance', *Urban Studies* 48(10): 1997–2011.

Beebeejaun, Yasminah, and David Vanderhoven (2010) 'Informalizing participation: insights from Chicago and Johannesburg', *Planning Practice and Research* 25(3): 283–296.

Boeije, Hennie R. (2009) *Analysis in Qualitative Research*, Los Angeles: Sage.

Brownill, Sue, and Juliet Carpenter (2007) 'Increasing participation in planning: emergent experiences of the reformed planning system in England', *Planning Practice and Research* 22(4): 619–634.

CAG Consultants; Carey Doyle, Tim Maiden, Richard Gale, and Andrew Rogers (2017) *Faith Groups and Meeting Places: Evidence Based Study*, London Borough of Barking and Dagenham.

Cooper, Brittany (2015) 'Intersectionality' in Lisa Disch and Mary Hawkesworth (eds) *The Oxford Handbook of Feminist Theory*, New York: Oxford University Press, 385–406.

Doyle, Carey (2018) *Migration, Planning and Superdiversity in Rural Northern Ireland and England*, PhD Thesis, Belfast: Queen's University Belfast.

Dwyer, Claire, David Gilbert, and Bindi Shah (2013) 'Faith and suburbia: secularisation, modernity and the changing geographies of religion in London's suburbs', *Transactions of the Institute of British Geographers* 38(3): 403–419.

Ellis, Hugh (2015) 'The re-creation of social town planning?', *Planning Theory and Practice* 16(3): 436–440.

Fainstein, Susan S. (2000) 'New directions in planning theory', *Urban Affairs Review* 35(4): 451–478.

Fincher, Ruth, and Kurt Iveson (2008) *Planning and Diversity in the City: Redistribution, Recognition and Encounter*, Basingstoke: Palgrave Macmillan.

Fincher, Ruth, Kurt Iveson, Helga Leitner, and Valerie Preston (2014) 'Planning in the multicultural city: celebrating diversity or reinforcing difference?' *Progress in Planning* 92: 1–55.

Fox-Rogers, Linda, and Enda Murphy (2014) 'Informal strategies of power in the local planning system', *Planning Theory* 13(3): 244–268.

Gaffikin, Frank, and Mike Morrissey (2011) *Planning in Divided Cities: Collaborative Shaping of Contested Space*. Chichester: Blackwell.

Gaffikin, Frank, Chris Karelse, Mike Morrisey, Clare Mulholland, and Ken Sterrett (2016) *Making Space for Each Other: Civic Place-Making in a Divided Society*, Belfast: Queen's University Belfast.

Gale, Richard (2005) 'Representing the city: mosques and the planning process in Birmingham', *Journal of Ethnic and Migration Studies* 31(6): 1161–1179.

Gale, Richard (2008) 'Locating religion in urban planning: beyond "race" and ethnicity?', *Planning Practice and Research* 23(1): 19–39.

Gale, Richard, and Simon Naylor (2002) 'Religion, planning and the city: the spatial politics of ethnic minority expression in British cities and towns', *Ethnicities* 2(3): 387–409.

Greed, Clara (1994) *Women and Planning: Creating Gendered Realities*, London: Routledge.

Greed, Clara (2016a) *Christianity and Planning: Pentecostal Churches*, Occasional Paper 1, Bristol: University of the West of England, Faculty of Environment and Technology.

Greed, Clara (2016b) 'Religion and sustainable urban planning: "if you can't count it, or won't count it, it doesn't count"', *Sustainable Development* 24(3): 154–162.

Manouchehrifar, Babak (2018) 'Is Planning "Secular"? Rethinking religion, secularism, and planning', *Planning Theory and Practice* 19(5): 653–677.

McClymont, Katie (2015) 'Postsecular planning? The idea of municipal spirituality', *Planning Theory and Practice* 16(4): 535–554.

Murtagh, Brendan, Brian Graham, and Peter Shirlow (2008) 'Authenticity and stakeholder planning in the segregated city', *Progress in Planning*, Special Issue on Authenticity and Stakeholder Planning in the Segregated City, 69(2): 41–92.

O'Neill, Diane, and Brendan Murtagh (2017) 'Equality law, good relations and planning' in Stephen McKay and Michael Murray (eds) *Planning Law and Practice in Northern Ireland*. London: Routledge, 191–205.

Phillimore, Jenny (2013) 'Housing, home and neighbourhood renewal in the era of super-diversity: some lessons from the West Midlands', *Housing Studies* 28(5): 682–700.

Radaelli, Claudio M., and Anne C.M. Meuwese (2009) 'Better regulation in Europe: between public management and regulatory reform', *Public Administration* 87(3): 639–654.

Sandercock, Leonie (2003) *Cosmopolis II: Mongrel Cities in the 21st Century*, London: Continuum.

Shortall, Sally, and Ruth McAreavey (2017) 'Gender, migration and development: can advocacy groups be more of a hindrance than a help?' *Social Sciences* 6(2): 49.

Thomas, Huw (2000) *Race and Planning in the UK and the USA*, London: UCL Press.

Thomas, Huw, and Richard Gale (2015) 'Multicultural planning in twenty-first century Britain' in Michael A. Burayidi (ed.) *Cities and the Politics of Difference: Multiculturalism and Diversity in Urban Planning*. Toronto: University of Toronto Press, 109–31.

Uitermark, Justus, Ugo Rossi, and Henk Van Houtum (2005) 'Reinventing multiculturalism: urban citizenship and the negotiation of ethnic diversity in Amsterdam', *International Journal of Urban and Regional Research* 29(3): 622–640.

Vertovec, Steven (2007) 'Super-diversity and its implications', *Ethnic and Racial Studies* 30(6): 1024–1054.

15 Planning for religion

As if it really mattered

Clara Greed

Abstract

Having looked at the relationship between religion, gender, cities and planning in the various chapters, this chapter discusses the specific problems that confront faith groups in dealing with the UK planning system with particular reference to Pentecostal churches as an example, although many of the observations could be applied to any faith group experiencing problems with the planners. First the lack of policy guidance or direction at national level is outlined. Then the inconsistencies and inadequacy of planning law in regard to religious development is outlined. The implications of this lacuna are discussed in relation to the economic, social, spatial and environmental factors. A case study is provided of the difficulties that the KICC mega church has experienced from the planners, in trying to accommodate their enormous and diverse congregation in London. In order to help ameliorate the situation in the future, the last section, a 'Toolkit', is presented, comprising a list of questions that planning authorities need to ask themselves in order to raise their consciousness of religious issues, and to enable them to plan with greater understanding and awareness of the spatial needs of faith groups of all sorts.

Key words: development plans, planning law, Pentecostal churches, sustainability, environment, transport, ethnicity

The problem so far

Contributors to this book have set out a range of perspectives (some contradictory) on the intersection of gender, women, religion, spirituality, urban space inter alia. Various contributors have demonstrated that historically many cities under both Christianity and Islam were purposely designed to accommodate religious space and buildings (Chapters 4, 5, 7 and 8). But 'the problem' today is that the statutory planning system seldom recognises the importance of religion as a factor that impacts upon land use and urban development policy (Chapters 6, 9, 12, 13 and 14), or the enduring asset value and cultural capital provided by existing religious buildings for a wide range of community purposes (Chapters 4,

12 and 10). Granted some manifestations of spirituality and religion are personal, internalised and pietistic (especially for women) and do not necessarily require physical buildings for communal worship (aspects of Chapters 3, 5, 10 and 11). But the greatest challenge for the planners and the planning system itself is to deal with those manifestations of religion which *do* generate major spatial demands for places to meet, for example as a result of the dynamic growth of Pentecostal mega churches or the need for mosques to meet for Muslim immigrant communities (Chapters 6, 9 and 14). But there seems to be 'no space' for religion within the planning system. For example, Carey Doyle (Chapter 14) has highlighted the fact that there is no specific land-use or development category within planning law for religious buildings.

Therefore this chapter first looks, as a case study, at the relationship between the planning system and Pentecostal churches, although many of the issues are also experienced by other faith groups looking for larger premises too (Gale, 2008). I will relate the discussion to the two main aspects of the planning system, namely urban policy making which is enshrined within development plans and planning policy documents, and secondly the operation of the development control system (also known as development management) which deals with planning applications for specific types of development within the local area (Greed and Johnson, 2014: Chapter 3). This is followed by the presentation of a Religion-Mainstreaming Toolkit, adapted from our previous Gender Mainstreaming Toolkit (RTPI, 2003), that might enable planners to become more 'conscious' of religious factors in planning, and thus make better-informed decisions when dealing with applications from faith groups.

This chapter draws on my fuller research report (Greed, 2016a), where more examples may be found, not just on Pentecostals but on a range of other religious groups' experiences of dealing with the planners too. As to finding relevant planning cases, a web search was undertaken of all planning applications, appeals and case law that related to 'places of worship' including using key words such as church, gospel, faith and Pentecostal. I used www.compasssearch.co.uk, which in association with the Royal Town Planning Institute provides frequently updated reports on current cases and appeals. I also searched using DCS (Development Control Services Limited) which is a password-only professional planning law site. I looked for relevant articles in the planning press, such as in *Planning* which can be accessed at www.planningresource.co.uk (by searching for the date of publication and the key name of the person or place given in the text). I continue to search and observe the situation.

The levels and scope of planning

Inconsistency

There is a lack of national policy guidance on the place of religious buildings and faith groups within strategic urban land-use planning. The situation is not consistent or coherent across the nation with some local planning authorities taking a more positive view than others towards the 'same' sorts of applications. Planning applications from faith groups are not being given adequate value because 'religion' does not fit in the planners' terms of reference in a predominantly secular planning system. Applications are seen through the lens of standards related to non-spiritual land uses such as commercial, residential and leisure developments. In this section I will now look at the different levels of planning, international, EU, UK central government and local planning authority level for indications of faith-related issues. Planning law presents many challenges. In addition, there are other areas of law, regulation and control which affect the chances of faith groups getting planning permission, including building control, health and safety regulations, highways policy, environmental sustainability policies and equality laws.

The international level of planning

The United Nations, as manifest for example in its global Sustainable Development Goals (the SDGs) (Greed, 2016b), has put considerable emphasis upon equality and diversity issues including gender and ethnicity, but there is scant reference to religion. International NGOs (non-governmental organisations), including churches and religious charities, are more likely to be concerned with these matters (Freeman, 2012). But, secular ex-pat relief agencies and development organisations often fail to see the economic and social importance of religion in developing countries (Narayanan, 2015).

The environmental sustainability movement has become one of the most important issues globally, and has taken on almost religious characteristics, in seeking to save the planet from an untimely end because of global warming. However the original UN definition of sustainability as set out in the first UN Earth Summit in Rio in 1987 included social equality and economic well-being as well as environmental sustainability (Brundtland, 1987). This definition was more people orientated, and more open to taking into account people's religious beliefs. Sustainability policy requirements have been cascaded down from the UN level into each nation state's policy-making objectives. UK planners have tended to concentrate on the environmental component of sustainability as that fits well with their physical, land-use-based planning system. Sustainability policy continues to be hobbled by its people-less-ness (Dempsey et al., 2011).

UK central government level

There is a lack of national policy guidance on the place of religious buildings and faith groups within strategic urban land-use planning, or on how local planning authorities should deal with applications related to places of worship. Planners do not see religion as a strategic planning matter (like employment or residential development) in spite of it having a very tangible physical effect on the built environment in terms of land allocation, new community development, land-use allocation, zoning and new development. Planning applications from faith groups are not being given adequate value because 'religion' does not fit in the planners' terms of reference. Therefore applications for religious buildings are likely to be judged by inappropriate criteria more related to commercial development (Wickham, 2014).

Therefore religion, faith and belief do not figure strongly in official guidance from the Department for Communities and Local Government (DCLG), which, note, in 2018 was renamed the Ministry for Housing, Communities and Local Government (MHCLG). The government is more likely to discuss religion indirectly in relation to education, social policy and diversity, and particularly in relation to ethnic minority issues, not spatial planning matters. There are few studies of development plans in respect of religion. The planning consultancy CAG is the front runner, as it has undertaken several studies on the attitudes and policies of planners towards the needs of faith groups (CAG, 2013) in conjunction with the GLA (Greater London Authority) (one of the few authorities to include religion fully in its diversity agenda) (GLA, 2007). A major report on the planning system's approach to places of worship by CAG highlighted the fact that there is no national strategy or high-level governmental policy statement on this matter (CAG, 2008).

Most local planning authorities, at county, city and local district levels, do not even have a specific policy statement in their development plan documentation on places of worship. As with gender issues, religion falls into the 'strategic policy filter' (Reeves, 2002) whereby it is not mentioned at central government level, therefore local planning authorities see no reason to mention it at local level, and if someone asks about why there is no policy in the local plan they will be told it is the responsibility of central government so everyone passes the buck.

National planning law controls both new development and 'change of use' of existing buildings. Building uses are classified under the UCO (Use Classes Order). There is no special use class for places of worship. Rather they come under the designation (D1) 'non-residential institutions', which also includes exhibition halls, public libraries, universities and health facilities. Some ancillary religious buildings such as church halls may be ruled to come under the designation D2 for 'places of assembly and leisure', which includes bingo halls, dance halls, swimming baths and sports and leisure centres (Greed and Johnson, 2104: 49). However, as Andrea Mulkeen explains (Chapter 12), religion did once have its own use class under post-war reconstruction planning legislation but with increasing secularisation this was changed. In 1985, the UCO classifications were streamlined and 'modernised' and religion lost its separate category (Mulkeen, 2014: 23) under D2 (Greed with Johnson, 2014: 49). In addition to the planners, building inspectors, fire officers, highways departments and others have statutory control over different aspects of building construction, providing many more reasons to refuse permission for what may be seen as unfamiliar types of buildings. All these issues are of major concern to faith groups, and in late 2015 the Faith and Place Network produced a briefing paper which, with the support of the RTPI, it submitted to parliament (FAPN, 2015)

Planning at local government level

Planners are meant to be enablers, to encourage and manage urban regeneration and renewal, especially within deprived areas. Greater emphasis has been put upon 'localism', neighbourhood planning and community involvement. But, only a few community representatives are permitted to sit on the newly created neighbourhood forums, to represent on average several thousand people, so the chances of minority religious groups being represented are low (Greed and Johnson, 2014: 34–36). Indeed, planners seem rather unsure and edgy as to how to deal with churches, especially non-traditional ones. A previous chief executive of the London Borough of Croydon expressed uncertainty as to how to deal with the recent effects of African immigration (sic) in terms of church provision, although this in a borough whose inner-city areas have had a high level of ethnic minority population since the 1970s! (*Planning*, 25.02.12). Paradoxically, in spite of this admission, Croydon has had one of the better Supplementary Planning Guidance (SPG) (more recently known as Supplementary Planning Documents (SPD)) on church growth among local planning authorities (Croydon, 2009). SPG documents are policy statements that do not quite make it into the main body of the development plan, but need to be taken into account as additional ancillary guidance in determining a planning decision. Clearly such policy guidance would be much stronger if it were made a mandatory requirement of the main policy document

Planning applications for church development are generally dealt with on an individualised basis with little reference to the wider strategic, urban, regional and even national policy statements on the matter: because none exist. In contrast, applications for retail development are always perused in relation to higher-level policy considerations such as the extent of the catchment area, city-wide impact, traffic implications and local employment and regeneration considerations. Mulkeen notes that some standards did exist in the 1950s on the provision of churches in New Town neighbourhoods, but there are no planning requirements today as to how much land should be allocated for religious building within residential areas (Mulkeen, 2014). However, the Anglican Church is still at an advantage compared with newer types of churches, in that along with other long-established traditional denominations they have certain automatic rights to develop under the Ecclesiastical Exemption rules (DCMS, 2010).

There are five denominations in England that have special exemptions from many of the rigours of planning law and building control, namely, the Church of England (Anglican), Roman Catholics, Methodists, Baptists and the United Reformed Church. Church buildings are listed under urban conservation policy to highlight their importance, and comprise 45 per cent of all listed buildings but are not subject to the same listing regulations. But ecclesiastical law predates and takes precedence over modern planning law (this limitation to state control was established in 503AD and is still in force). Ecclesiastical exemption entitles the churches included to various 'permitted development' rights for example in relation to the benefice of the parish to build a vicarage and to undertake other forms of church-related development within the curtilage of the church building. In comparison many newer and non-traditional churches, and faith groups, have no right to develop new buildings or to change the use of existing ones.

Equality legislation

As touched upon by Carey Doyle (Chapter 14), the 2010 Equality Act lists seven protected categories that should be taken into account in all areas of policy making, including (in theory) urban planning. These are Age, Disability, Gender Reassignment, Pregnancy and Maternity, Race, Sex (gender) and Sexual Orientation, and Religion and Belief (but not social class note). Yet, social equality issues, including gender, have generally been given low priority compared with environmental considerations within the planning system (Greed, 2005; Reeves, 2005). Sexuality is increasingly given primary attention as the factor defining a person's identity, whereas religion has a low profile (Habermas and Ratzinger, 2007).

Yet, the Equality and Human Rights Commission seeks to integrate equality assessment into all aspects of policy making for all government departments, including planning, and the Public Sector Equality Duty (PSED) exists to ensure this happens (DCLG, 2011). The PSED, in respect of the needs of specific groups, is meant to be taken as a material consideration in determining planning applications and dealing with appeals, but this has not been widely enforced. Implementation of equality issues has been very limited in planning departments. Many planners imagine equality matters are nothing to do with them, owing to the technical and environmental nature of their work (Greed, 2005). Religion may be seen as simply a personal choice which is nothing to do with town planning. Planners have always had difficulty dealing with aspatial issues (Foley, 1964) (that is social, cultural, non-land-use matters) and find it hard to integrate them into physical spatial policy making (Dempsey et al., 2011; Buckingham, 2013).

Examples of the planners' attitudes to church-related planning applications

Explanation of categories

In this section I discuss the problems churches encounter in relation to planners' attitudes towards them, and the economic and social implications of this ignorance. I also include a section on the 'physical' architectural aspects of church growth, as much of development control is strongly linked to the building stock and how it is utilised. Then I provide a case study in the next section which will highlight the environmental sustainability and transport policy implications of not recognising the importance of church development within the planning system.

Economic issues

One of the most frequent reasons for the refusal of planning permission for new churches is because they are seen as taking up space that is allocated for employment and industry. Because of the lack of suitable available premises, such as redundant church buildings or empty public buildings, many new Black Pentecostal churches have resorted to applying to use empty industrial units, often on trading estates on the edge of town, where, to all accounts, there is little demand for occupation by businesses. Churches are generally viewed as an obstruction, as a negative factor that contributes nothing to urban regeneration

For example, application for a change of building use, from industrial to church use, in south London, was dismissed over 'employment fears' (reported 29.06.2012, DCS 100-077-056). Another church failed to overturn an enforcement notice against its use of an industrial unit in southeast London as a church, the planning inspector stating that the community benefit did not outweigh the conflict with employment policies, although several other similar industrial buildings were currently empty (29.01.2010, DCS 100-065-817). Even if churches do get planning permission they are not necessarily free of the economic clutches of the planners. For example, the Christ Apostolic Church Mount Zion in the London Borough of Southwark won its appeal against being required to pay a new planning levy being imposed on the development, which would usually be raised on commercial property, which seemed totally disproportionate for a small and poor faith group (*Planning*, 30.11.12). No doubt the appeal was won through the power of prayer and the intercession of God rather than through planning law. In response to these problems, there is a significant and powerful minority of planning consultants and advocates who seek to help Black Pentecostal churches, and some of them are themselves of Afro-Caribbean and African origin, for example the Adebayos who run TA Planning Consultants.

The planners still seem to have an old-fashioned attitude to what constitutes the economy, and what might create growth and urban regeneration, in which 'industry' is still given a privileged position. They do not see churches as a source of employment, regeneration or indeed of creativity, although 'creative industries' and entrepreneurial activity may arise from within the church community (Stott, 2012). But 'church' is not seen as an engine of the economy (as Patrick Anderson a planner working at Planning Aid for London has discussed with me). Thus religious capital is not recognised as a change agent, or as a force as potentially powerful as economic capital, because it does not 'fit' into the planners' view of how the world should be. Patrick explained to me that while the globalisation of economies has actually undermined English industry – hence so many empty industrial units – it has also created increased immigration and thus the demand for places of worship, a role which the industrial units could readily fulfil. Some planners still live in a world in which they sincerely believe that one day 'industry' will re-emerge so they have to save the industrial units for that time.

Social benefits

There are many social benefits to permitting new church development. Black Pentecostal churches often provide day care facilities, education and careers guidance and undertake a host of charitable activities and positive interventions in the surrounding neighbourhood, all of which actually contributes to residents finding work and having the housing, education and childcare to participate in the workforce. But urban regeneration is so obsessed with employment and investment issues, and increasingly with achieving environmental sustainability and 'green' credentials, that often it forgets all about the essential social support system needed to enable people, especially women, to get into the workforce (Greed, 2005).

Some white churches may occasionally experience some of the same problems in getting planning permission. But it would seem they are often treated more favourably as both class and ethnicity modify attitudes towards religion. Large predominantly white evangelical churches seemed to be viewed as highly respectable and part of the 'establishment' by some planners, even if they are charismatic in their form of worship and beliefs (the polite white middle-class version of Pentecostalism). Such churches may also run educational facilities, housing associations, charities, nurseries and almost always some form of coffee bar or drop-in centre and an Alpha course (Hunt, 2004). Such white evangelical and charismatic churches are increasingly involved in social action, reform and local politics. Successful planning applications include a new central London office development for OASIS (a major evangelical organisation) as reported in the popular evangelical magazine *Christianity* (June 2008: 9) (see www.christianitymagazine.co.uk). Birmingham City Council willingly granted permission for the conversion of a factory building in Frankfort Street, Newtown to a church with educational facilities run by New Creation Ministries, which operates from St Cyprian's Memorial Church (*Planning*, 05.11.10). Likewise in leafy Hampshire, church use restrictions were actually relaxed to enable a new church to provide youth group meetings and other associated activities on its premises, but this was a white church (01.07.2011, DCS 100-072-266. In previous research I found that there seemed to be less concern with parking controls at traditional suburban churches (Greed, 2012).

Many planners and local councillors with a secular background seem to completely misunderstand the beliefs of faith groups (and appear religiously illiterate) and treat them with suspicion as likely to be homophobic and out of step with the equalities agenda. The public expression of one's beliefs, and free speech, is being increasingly curtailed. The 2015 Counter Terrorism and Security Act implies that if a person has strong religious or political beliefs this might indicate links with terrorism, although this is more likely to be related to Muslim than Christian groups (www.gov.uk/government/publications/prevent-duty-guidance) but given the racism and the apparent visual inability of some local authority officers to distinguish between the different types of 'people of colour' this trend can result in suspicion of all Black and ethnic minority people. Therefore some Black churches will employ an 'agent' (usually a white planning consultant) to submit their planning application and to meet with the planners to increase their chances of success.

Architecture and building stock issues

While established church congregations are generally tolerated by the planners, what planners like even more is empty, cold, redundant churches which are to be commended for their age and quality of architecture. In line with the widely held false belief that the church is in decline, empty church buildings are seen as a wonderful opportunity for re-use as community centres, apartments, art galleries, and even bars and clubs, within the context of regenerating the inner city. Some older Protestant churches are complicit in this process, glad to off-load their old expensive properties and heating costs. But Roman Catholic churches are less likely to experience this fate because once 'consecrated' their holy status is not easily rescinded. Some 'trendy evangelical vicars' may imagine that turning their church into a community centre, and thus removing any sign of ecclesiastical imagery, will improve their chances of bringing people in (but removing the building's religious trappings reduces the chance of future use for weddings or other formal religious ceremonies). But one should at least be grateful that the planners *do* value church buildings nowadays, even if for the wrong reasons. Anthony Fyson notes, in the 1970s in the days of cost–benefit analysis calculations, a priceless medieval church, which was in the way of a new airport, was valued as only being worth 10,000 pounds (*Planning*, 17.06.11). Nowadays the architectural, historical and touristic value of churches is valued much more highly than their religious, spiritual, social and community usefulness. But, it is estimated that the good work done by churches and faith groups in society is worth at least 3 billion pounds a year (Cinnamon Network, 2015). For example, the development agency responsible for regional development in the northwest of England calculated that faith groups contributed £95 million of voluntary work to the region each year (Atherton et al., 2011: 39).

The emphasis on architectural purity can crush church growth. Some still-lively, expanding churches, located in historical buildings, have found attempts to expand and build on extensions to be thwarted by the planners. A three-storey extension proposal for a Victorian church in east London was deemed to destroy the symmetry of this listed Bath stone building (DCS 100-074-444). But the need for an extension should have been seen as a sign of a healthy, growing church and encouraged. Applications for extensions are a particular problem for erstwhile poorly attended Roman Catholic churches which have experienced unprecedented growth in recent years as a result of Polish immigration to the UK.

In spite of a huge demand for church premises from new church groups, including the Pentecostals, one seldom finds an old redundant church being made available to new faith groups. For example St Paul's Church in Bristol, built in 1790 was restored through the Heritage Lottery Fund, and was further adapted in 2005 to make space for a circus arts academy. Note this is a majority ethnic minority area where just up the road one can find Black churches precariously holding on to tenancy of industrial units who would have jumped at the chance to be offered such redundant church buildings in the area. But maybe this building was seen as too posh to give to the local residents? This is one of the issues raised by the following case study.

Case study: the KICC church

Cinema sagas: insiders and outsiders

To illustrate the issues, I will give a summary of the KICC church saga, which I have been following for several years, having previously covered this church's troubles in other work (Greed, 2011). This case study also looks at the environmental sustainability, public health concerns and transport planning issues involved. KICC (Kingsway International Christian Centre) is a huge church of over 12,000 members from 46 different nationalities, which had been seeking to build a mega church which would seat more people than any existing cathedral in Western Europe. The church had to move from its previous premises to make way for the London Olympics with a promise from the planners that any alternative site would be looked upon favourably. But this promise was not kept. Subsequently the church has put in many applications on a range of sites and buildings within London. For example they bought a redundant cinema in Bromley, inner London, near to Crystal Palace, and started using it as a church, in the hope of receiving retrospective planning permission. They were subject to the threat of enforcement proceedings on a whole series of technicalities and experienced a number of negative appeal decisions.

'Who' comprises the local community?

Long-established local residents expressed objections over a range of issues, many of which were actually quite racist in tone. In particular they expressed concern about 'outsiders' taking over the building, thus apparently depriving existing residents of its use. Fears were expressed by some local residents about noise, disturbance, increased crime and anti-social behaviour taking place, and overcrowding on the pavement outside. Reports were made about church goers spitting and urinating on the pavement outside (www.crystal-palace-mag.co.uk). The locality has a substantial and long-established ethnic minority population and members of the KICC church are known for their righteous living and good works, and are hardly part of the perceived criminal fringe. In the end the cinema building was turned into an arts complex, a use which arguably does not reflect the needs of the majority of the local population but that of the metropolitan elite. But culture, media and the arts are far more likely to be seen by the planners as drivers of urban regeneration and the local economy than religion.

Whose noise and disturbance?

Not only does this example raise the question of 'who' comprises 'community', and what 'local culture' consists of, in the minds of the planners, but it also raises issues as to what local planning authorities see as valid leisure, entertainment and cultural uses. Many local authorities are keen to promote the 'night time economy' and the 24-hour city, as explained in Gill Reeves' chapter. Such policies usually result in over-provision of licensed drinking premises, with attendant noise, public drunkenness and street urination. There is clearly a double standard as many churches are refused permission on the basis of noise and disturbance, although many congregations are likely to be teetotal and careful not to disturb the peace. In addition to 'noise', many a church has been refused permission because of a range of technicalities regarding noise generation, but also building design, car parking and public health issues. Noise is clearly a subjective factor judged according to the culture, class and race of the noise makers (*Planning*, 4.12.09). There have been several cases where expensive sound insulation, double, if not treble glazing, and restrictions on times and length of church meetings have been introduced. The whole planning process can be dragged out on technicalities, arguably to deter applicants persevering. The KICC church was keen to establish a series of satellite churches throughout London and in every instance such issues were likely to be raised to slow down or prevent their progress.

Environmental and traffic issues

Environmental sustainability is a primary concern to modern planners, and this usually results in policies concerned with controlling car usage, improving public transport, and reducing the outward growth of cities to reduce long commuting journeys (Greed with Johnson, 2014). When the KICC church appealed about not being allowed to use the cinema building (discussed above), the planning inspector had raised questions regarding 'who' is entitled to use the pavement outside, and had expressed concerns about the church generating too many pedestrians, because many of the congregation would walk to church, as well as making a range of curious comments about the use of bicycles, local taxis and public transport by church members overwhelming the capacity of the local infrastructure. When the building was re-opened as a cinema presumably that too generated lots of people, but perhaps they are seen as 'the right kind of people'?

Back in 2009, KICC applied for an 8,000-seat mega church in Rainham, Essex (on the outskirts of London), on an industrial estate. KICC partly chose this site because there were no near neighbours so they could not be accused of causing a disturbance and being too noisy. This application was refused on the grounds that the site was already allocated for industry, although most of the units remained empty. Another planning inspector dealing with an appeal regarding the failed planning application for the church to be located in Rainham, actually stated that the church would put too much demand on local bus services (isn't that what buses are for?). This is extraordinary in that nowadays transportation planners are obsessed with getting people out of their cars and on to public transport for 'green' environmental sustainability reasons. So clearly there are times when the priorities of the sustainability agenda are suspended because of a greater concern with controlling religious development. Planners would not apply the same attitudes to office or retail development, where they are encouraging people to use public transport and leave their cars at home, by means of strict controls on parking and often the provision of improved public transport. Church leaders have commented that if the KICC church (which regularly attracts thousands) had been a major retail developer (such as the Westfield shopping centre at Shepherds Bush), or a football team (such as Chelsea), which both attract thousands of supporters, the application would have been dealt with in a very different way, with major changes to local transport policy to enable fans to reach the stadia each weekend.

The KICC church tried and tried to get planning permission on several different buildings and new sites over the years, both inner city ones such as the cinema, and edge-of-city sites on industrial estates. Meanwhile the congregation remained fragmented and reduced to holding multiple services in a range of scattered satellite venues. Eventually, the KICC saga was resolved by the church 'leapfrogging the green belt' and moving into more rural premises: Buckmore Park, Maidstone Road, Chatham, to the east of London, as well as still retaining churches in Walthamstow and other satellite locations (see www.kicc.org.uk). People are ferried to and from the new Chatham site by minibus or commute by car, as public transport is not direct, and arguably this has had more impact on the environment than just letting them stay in London, and further undermines the environmental objectives of the Greater London Development Plan.

Conclusion

For religion to be valued and taken seriously within the planning process, there needs to be a change in the world view of planners, in terms of the subcultural values of the profession which are mainly secular, and concerned chiefly with spatial and quantitative factors (Greed, 2013; Fewings, 2010). Religious buildings need to be treated like any other land use, such as retail or commercial development. This will require doing a national survey of building stock and also levels of demand, in order to develop spatial policies which provide land allocation and building stock policies to accommodate growing demand.

Churches are not just historical cultural artefacts but living parts of the community. As John Betjeman foretold in his poem 'The Town Clerk's Views', in

1948, 'Every old cathedral that you enter by then will be an Area Culture Centre' (this being highlighted in the Preface to Victor Moore's book on *Planning Law*) (Betjeman, 2006: 146; Moore, 2010). The economic, social, spiritual and spatial importance of faith buildings needs to be acknowledged by the planners, and ideally the issue of religion needs to be introduced into university planning degrees and continuing professional development courses. At the detailed level, there needs to be a major overhaul of planning law in respect of the needs of faith groups. Religious land use needs to be recognised as an important and distinct category of land use and building occupation with the Use Classes Order (UCO).

But how can we generate a change in the planners' attitude towards religion? Drawing on methodologies used in work on 'women and planning' (another marginalised and under-valued topic) in the following section I have adapted the Gender Mainstreaming Toolkit we originally produced for the RTPI in 2003 to create a new toolkit that might help planners mainstream religion and faith issues into their work. It is basically a means of raising awareness and consciousness of these issues at each stage of the planning process (RTPI, 2003). My prayer for the future is that religion will be recognised and taken into account by the planners as an important factor within society and urban space. But I also plead that religious organisations will fully recognise women as human beings created in the image of God.

Proposed religion and planning toolkit

Urban policy makers and planners should ask themselves these questions

1 What resources and funding are available in the planning department to investigate the levels of demand for religious buildings?
2 Does the policy team include planners who are knowledgeable about religious issues? What additional training is available to increase their knowledge?
3 Who are perceived to be the planned? Which religious groups are included? Are community representatives available for public participation? Do they include a balance in terms of ethnicity, gender, locality and religious affiliation? Is training about the planning system available to religious representatives?
4 Has a survey been undertaken of the amount of, and demand for religious building stock in the borough? Do you have up-to-date statistics on demand, refusals for new religious buildings or change of use and reasons given?
5 Do the objectives of your development strategies and local plans take into account demand for religious buildings? Do any of your other main policy topics, such as employment, transport, housing, have implications for religious development?

6 How is the plan evaluated? What are your criteria of success? Is it based on economic, social, cultural or religious factors?
7 How do your policies regarding the location and accommodation of religious demand relate to achieving environmental sustainability? In particular does your transport policy accommodate and increase accessibility for church goers including pedestrians and users of public transport?
8 How is the policy implemented? Are churches and other faith groups still involved in ongoing monitoring or have they been dropped once development has taken place?

References

Atherton, J., Baker, C. and Reader, J. (2011) *Christianity and the New Social Order: A Manifesto for Fairer Futures*, London: SPCK.

Betjeman, J. (2006) 'The Town Clerk's Views' in *John Betjeman: Collected Poems*, London: John Murray, 144–147.

Brundtland (1987) *Our Common Future*, World Commission on the Environment and Development, Oxford: Oxford University Press.

Buckingham, S. (2013) 'Gender, sustainability and the urban environment' in I. Sánchez de Madariaga and M. Roberts (eds) *Fair Shared Cities*, London: Ashgate, 21–32.

CAG (2008) *Responding to the Needs of Faith Communities: Places of Worship: Final Report*, London: CAG, Cooperative Advisory Group Planning Consultants.

CAG (2013) *Northampton Faith Communities Profile and Places of Worship Audit and Needs Assessment*, June, London: CAG.

Cinnamon Network (2015) *Cinnamon Faith Action Audit Report*, London: Cinnamon Network. www.cinnamonnetwork.co.uk/.

Croydon (2009) *Places of Worship: Supplementary Planning Guidance, No. 9*, Croydon: London Borough of Croydon.

DCLG (2011) *Public Sector Equality Duty (PSED)*, London: DCLG and EHRC (Equality and Human Rights Commission).

DCMS (2010) *The Operation of the Ecclesiastical Exemption and Related Planning Matters for Places of Worship in England*, London: Department of Culture, Media and Sport (DCMS).

Dempsey, N., Bramley, G., Power, S. and Brown, C. (2011) 'The social dimension of sustainable development: defining urban social sustainability', *Sustainability and Development* 19: 289–300.

FAPN (2015) *Faith Groups and the Planning System: Policy Briefing*, Roehampton University: Faith and Place Network, written by Andrew Rogers and Richard Gale. www.roehampton.ac.uk/Courses/Humanities/Calendar-Humanities/Launch-of-the-Faith-and-Place-network-policy-briefing/.

Fewings, P. (2008) *Ethics for the Built Environment*, London: Taylor and Francis.

Foley, D. (1964) 'An approach to urban metropolitan structure' in M. Webber, J. Dyckman, D. Foley, A. Guttenberg, W. Wheaton and C.B. Wurster, *Explorations into Urban Structure*, Philadelphia, PA: University of Pennsylvania Press, 21–78.

Freeman, D. (2012) *Pentecostalism and Development: Churches, NGOs and Social Change in Africa*, Jerusalem: The Van Leer Jerusalem Institute.

Gale, R. (2008) 'Locating religion in urban planning: beyond race and ethnicity', *Planning Practice and Research* 23(1): 19–39.

GLA (2007) *Planning for Equality and Diversity in London*, The London Mayor's Supplementary Guidance to the Greater London Strategic Development Plan, London: Greater London Authority.

Greed, C. (2005) 'Overcoming the factors inhibiting the mainstreaming of gender into spatial planning policy in the United Kingdom', *Urban Studies* 42(4): 1–31.

Greed, C. (2011) 'A feminist critique of the post secular city: god and gender' in J. Beaumont and C. Baker (eds) *Post-Secular Cities: Space, Theory and Practice*, London: Continuum, 104–119.

Greed, C. (2012) 'Planning for sustainable transport or for people's needs', *Urban Design and Planning Journal* 165(4): 219–229.

Greed, C. (2013) 'A feminist perspective on planning cultures: tacit gendered assumptions in a taciturn profession' in G. Young and D. Stevenson (eds) *The Ashgate Research Companion to Planning and Culture*, Farnham: Ashgate, 89–104.

Greed, C. (2016a) *Christianity and Planning: Pentecostal Churches*, Occasional Paper 1, Bristol: University of the West of England, Faculty of Environment and Technology.

Greed, C. (2016b) 'Religion and sustainable urban planning: "if you can't count it, or won't count it, it doesn't count"', *Journal of Sustainable Development* 24: 154–162.

Greed, C. with Johnson, D. (2014) *Planning in the United Kingdom: An Introduction*, London: Palgrave Macmillan.

Habermas, J. and Ratzinger, J. (2007) *The Dialectics of Secularisation: On Reason and Religion*, Rome: Ignatius Press.

Hunt, S. (2004) *The Alpha Initiative: Evangelism in the Post-Christian Era*, Farnham: Ashgate.

Moore, V. (2010) *A Practical Approach to Planning Law*, Oxford: Oxford University Press.

Mulkeen, A. (2014) *The Planning System and Places of Worship: The Extent to Which the Planning System since the Second World War Has Facilitated the Needs of Faith Communities for Places of Worship*, MA in Urban and Regional Policy, London: University of Westminster.

Narayanan, Y. (ed.) (2015) *Religion and Urbanism: Reconceptualising Sustainable Cities for South Asia*, London: Routledge.

Reeves, D. (2002) 'Mainstreaming gender equality: an examination of the gender sensitivity of strategic planning in Great Britain', *Town Planning Review* 73(2): 197–214.

Reeves, D. (2005) *Planning for Diversity: Planning and Policy in a World of Difference*, London: Routledge.

RTPI (2003) *Gender Mainstreaming Toolkit*, London: Royal Town Planning Institute, authors D. Reeves, and C. Greed and team. www.rtpi.org.uk.

Stott, V. (2012) 'Creative space: sacred space', *Movement* (London: Student Christian Movement Magazine) 142(Autumn): 14–15.

Wickham, R. (2014) 'Aspects of growth and change', Paper presented at *Faith and Place Perspectives Conference*, Royal Foundation of St Catherine, 4 December (presumably available as a video of his talk on their network).

Name Index

Page numbers in **bold** denote tables

Subject Index

For Product Safety Concerns and Information please contact our EU
representative GPSR@taylorandfrancis.com
Taylor & Francis Verlag GmbH, Kaufingerstraße 24, 80331 München, Germany

www.ingramcontent.com/pod-product-compliance
Ingram Content Group UK Ltd.
Pitfield, Milton Keynes, MK11 3LW, UK
UKHW020012200926
475702UK00014B/1838

* 9 7 8 1 0 3 2 0 8 5 3 4 0 *